THE SEARCH
FOR
SASQUATCH

THE SEARCH FOR SASQUATCH

FOR

SASQUATCH

BY

LAURA KRANTZ

A WILD THING BOOK

Abrams Books for Young Readers
New York

Cataloging-in-Publication Data has been applied for
and may be obtained from the Library of Congress.

ISBN 978-1-4197-5818-8

Text © 2022 Laura Krantz
Illustrations © 2022 Abrams Books for Young Readers
Edited by Howard W. Reeves
Illustrations and book design by Rafael Nobre

Printed and bound in China
10 9 8 7 6 5 4 3 2 1

Abrams Books for Young Readers are available at special discounts when
purchased in quantity for premiums and promotions as well as fundraising
or educational use. Special editions can also be created to specification.
For details, contact specialsales@abramsbooks.com or the address below.

ABRAMS The Art of Books
195 Broadway, New York, NY 10007
abramsbooks.com

To Grover
And to my parents, Chip and Louise Krantz,
who told me to go outside and play

CONTENTS

NESTS

Skreek . . . Skreeeeek . . . Skreeeeeeeeeeeek

Whiplike, bendy branches scraped down the sides of our giant black pickup truck, like a witch's fingernails. I tried not to lose my lunch as we bounced along a bumpy gravel logging road in the hinterlands of Washington's Olympic Peninsula. Just as I rolled down the window to get some fresh air, Shane Corson, my bearded, mountain man guide, pulled to a stop in front of a padlocked steel gate. I looked around, not sure if this was the right place. We were in the middle of nowhere! But Shane killed the engine, hopped out of the truck, and fished a key from his pocket. The giant lock popped open, and Shane loosened the chain so he could swing the gate wide. This was it! I suddenly got so excited that I forgot how barfy I'd just felt. We were about to go deep into a patch of tangled woods and towering spruce trees to see something that was off-limits to, well, just about everyone.

Shane hopped back into the driver's seat sporting an ear-to-ear grin. "It's not much further up the road," he said, clearly feeling as excited as I was about what he planned to show me. We inched forward through the thickening underbrush, and a few bumpy minutes later, we had gone as far as the truck could take us. We had to hoof it from there.

I'd met Shane about a year ago, on a camping trip in Oregon. But I didn't really get to know him until a few months later because that first time, he wasn't too excited about talking to me. Why? Well, let me introduce myself. My name is Laura, and I'm a journalist. My job is asking lots of questions and writing about what I learn for others to read. Sometimes that can make the people I talk to pretty nervous, especially if they think someone is going to make fun of them and their ideas and beliefs. And Shane? I think he felt particularly worried because of one very specific interest: Bigfoot.

That's right. I said Bigfoot. Sasquatch himself.

Now, if you've never heard of Bigfoot before, picture an enormous, apelike creature that's ten feet tall and might weigh as much as one thousand pounds. It's covered in hair, walks on two feet (like us), and leaves giant footprints behind—or, at least, that's what we *think* it looks like. There are lots of fascinating and terrifying stories about Bigfoot from people who claim to have seen or heard it, although no one has ever been able to show any real, scientifically acceptable proof. But Shane swore that what was hidden out there in the woods could help make the case for Bigfoot.

I jumped down from the truck, perfumed myself with big spritzes of bug spray, and tightened the laces on my hiking boots. Then I straightened up and followed Shane into the wall of shrubs and blooming rhododendron bushes. As we plunged through the underbrush, Shane kept a brisk pace. He clearly knew where he was going, while I seriously struggled to keep up. Behind me, the truck disappeared.

Jeez, I thought to myself. *I hope I don't have to find my own way back.*

Huckleberry brambles caught on my clothes and left red, raised scratches down my arm. It all looked so beautiful and wild, but I didn't have any time to take in the scenery.

Then Shane disappeared down a steep slope, and I lost sight of him for a second. I stopped still—and it was dead quiet. I couldn't even hear him moving anymore. Just as I started to worry, I pushed through another thick wall of shrubs and found him standing quietly at the base of some trees. He looked at me and then looked down.

"Here we are. This is it," he drawled, ho-hum, like it was no big deal. He moved a little bit farther and gestured at the ground. I looked to where he was pointing, and my jaw dropped. If I had been trying to play the part of the calm and unflappable reporter, I'd just failed.

"Whoa!" I exclaimed, not at all professionally.

This wasn't what I had expected—but it was definitely what I'd hoped for.

"This is . . . this is **CRAZY.**"

We stood in front of a pile of intertwined sticks and branches, woven together so carefully that they looked like they'd been made into a giant nest. It was at least eight feet across—so big that I could have comfortably lain down in it. I could even have stretched out. In fact, Shane had already done this himself, when he came out here before.

"It felt like a mattress," he said with a grin. "I felt small. Very small."

It truly looked like a bird's nest. It could have *been* a bird's nest. But, of course, as far as I knew, no bird on the planet made nests that size.

And that wasn't the only nest, either. Shane pointed out six others nearby, hidden between clumps of trees, with a few small ones tucked

into low branches. He said there were others, too, but farther away and a little harder to get to.

"You would have been amazed when we first came down here," he said, obviously pleased with my reaction. "Three years ago, when we first saw these, they looked even better then. These weren't just slapped together." The *we* he referred to was the Olympic Project—a Bigfoot research group Shane belongs to. A few years ago, the man who owns this land found the nests when he was out inspecting the trees on his property. They confused him, so he asked the Olympic Project to come take a look at them—twenty-one in all—to see if they could puzzle out what had made them. Shane, who had spent his life hunting and fishing and camping, knew that these nest things were really unusual.

"It's not a bear bed or an elk bed or a deer bed." He scraped at a tree, pulling off bits of loose bark, and gathering it, along with dead leaves, small branches, and other debris, into a loose, sloppy pile. "That's what a bear bed looks like. But these are nests, and in all my years I've never come across anything like this," he said. "They look more like gorilla nests than anything else."

Gorillas make nests? I asked myself.

Later, I looked up "gorilla nests" online, and Shane was right: Gorillas do build giant nests, similar to this one. But there aren't any gorillas in this part of the world. Before I could ask more questions, Shane took off down the trail again, yelling back over his shoulder for me to keep up. Five minutes later, he screeched to a halt. He scanned the woods, like he was lost.

Oh no—we're lost and some giant nest-building thing is out here with us, I thought to myself.

Then he wrapped his hand around the branch of a nearby huckleberry bush and showed me the end of it. It had been broken off. I looked at him blankly—I didn't get it. Then he showed me another branch. And another. And another. All the same.

"These huckleberry branches have all been snapped and their leaves stripped clean to make these nests. There are no teeth marks. Something had to have hands to snap it off. Strong hands, because some of these branches are a couple of inches thick. You can't just break something like that in half," he pointed out. I gave it a try. It would have taken someone (or something!) much stronger than I was to do it. And that nest we just saw was piled high with those branches.

What did it all add up to? Something with the strength to break thick branches. Something that makes nests. Something that likes the solitude of the woods. So I was not the teeniest bit surprised when Shane finally said what we'd both been thinking: that *maybe*—and it was a BIG maybe—Bigfoot made these nests.

BIGFOOT VS. SASQUATCH

The names "Sasquatch" and "Bigfoot" refer to the same creature. We think the word Sasquatch comes from the Coast

Salish—a group of Indigenous peoples (Native Americans, as they're known in the United States, and First Nations, as they're known in Canada) who live along the Pacific coastline from British Columbia, Canada, down to Oregon. The word they used was *Sasq'ets*, which means "wild man" or "hairy man." A white man named J. W. Burns Anglicized the name, meaning he adapted the Indigenous word into English.

The word Bigfoot appeared for the first time in 1958, when a group of loggers in Northern California found giant footprints in the mud near their work site. They began to use the name "Big Foot" to talk about the creature that left the prints and also moved their equipment. A reporter named Andrew Genzoli wrote up a story about the loggers and their new mysterious friend, calling it "Bigfoot," and the name stuck.

"We're not saying it's definitely Sasquatch," Shane said. "We're saying we don't know what made these nests. They're unknown nests, with unknown hair mixed in with the foliage, and unknown animal behavior behind it. The goal is to get to the bottom of the mystery."

I knew Shane leaned toward the possibility of Bigfoot. But he wanted to be careful. For starters, they still didn't have the right kind of evidence to prove it. And Shane also knew that most of the world thought Bigfoot was nothing more than a big joke, a myth. He'd been laughed at by enough people that he didn't tell just anyone about his interest. And even though I wasn't sure Bigfoot was real, I had started to understand how people ended up tromping around in the woods looking for evidence.

Then Shane said something that was super important and reminded me *why* I came all the way out here, into the middle of nowhere. "Something *made* these. Maybe we can't say for sure that Bigfoot made them, but these things are weird. Why wouldn't we want to ask questions?"

Asking questions—that's the point of all this. Because I was standing in front of something I couldn't explain, and while part of me felt a little silly about being there, I mostly just wanted to find out more. I mean, I'm a journalist—it's kind of my job. But I also felt really curious. What is Bigfoot? What do we know about this creature? Why are there so many sightings and so little proof? And why are all the photos that we do have so blurry? I had no idea if I'd find Sasquatch, or, well, some other explanation, but I wanted to get to the bottom of the Bigfoot phenomenon.

While we had been talking, we'd hiked back to the nest site. Shane bent over and looked more closely at a section of the nest where a wedge, like a big, fat slice of pizza, had been removed. I wasn't the only visitor he'd brought out here. A few months back, a scientist came to collect samples of the nest, hoping to gather clues about this mystery in the woods. Clearly a lot of other people also had questions, and I wasn't sure any of us knew what we were going to find.

GROVER

For most of my life, I didn't think there was any way that Bigfoot could be real. It was just a story, a myth, a campfire tale. Sure, I'd heard about Bigfoot as a kid. I'd watched that cheesy old movie *Harry and the Hendersons,* about the family that hit Bigfoot with their car and then brought him to live with them. I'd seen the silly Bigfoot headlines in the tabloid newspapers while waiting in line at the grocery store. But I didn't *really* think about Bigfoot, and if I did, it was mostly as a Big Joke. So how on earth did I end up in the middle of a thick forest, with a Sasquatch seeker, gawking at what might be a giant nest made by Bigfoot?

Well, I can pinpoint the exact moment that I started thinking more about Bigfoot. I used to work in Washington, DC, which is a fun city with lots of amazing (and free!) museums. My personal favorite is the Smithsonian's National Museum of Natural History, and one weekend, I took a trip to go check out the cool rocks and dinosaur fossils. The museum had just opened a brand-new exhibit, called *Written in Bone.*

Anthropologists—scientists who study how humans have changed over time—use ancient skeletons to learn more about the lives of our ancestors, and this exhibit had all kinds of old bones on display. It turns out bones can tell us a lot about the people who came before us. Like, bones that have teeth marks in them, from leopards and saber-toothed tigers? Well, those show how some of our ancestors might have met an untimely end. Scientists can also analyze the minerals in bones to see what these ancient ancestors ate and how healthy they were. It's really cool stuff, and I was totally fascinated.

WHAT IS ANTHROPOLOGY?

We're going to hear from a lot of anthropologists in this book. But what is anthropology, exactly? Anthropology is the study of human beings, both in the present and in the past. *Anthro* means "human," and *ology* means "to study." There are several different types of anthropology, but the two main ones that we'll learn about in this book are physical or biological anthropology and cultural anthropology.

Physical anthropology looks at how humans evolved from other animals and how they have adapted to all the different environments they've lived in. Physical anthropologists do this by studying humans (both living and dead), as well as the ancient remains of our human ancestors. They also study monkeys and apes—primates—because scientists think we all came from a common relative millions of years ago (we'll learn more about this in the next chapter). Cultural anthropology looks at how people live—the rules they follow and what values and ideas they think are important. Depending on where you live and how you grew up, you may have very different ways of looking at the world. Cultural anthropologists often try to learn more about how different groups think by living with them and seeing what their daily lives are like.

At the very end of the exhibit, in a big glass case, I saw a tall human skeleton, leaning back. The skeleton of a giant dog stood in front of it on its hind legs, with its front paws on the human's shoulders. Next to the display case was a photograph of the man and the dog hugging, back when they were both alive. The people at the museum had wired the two skeletons back together and set them up to look just like the image. The caption said the man had been a professor of anthropology at Washington State University and had left his remains and those of his beloved Irish wolfhound Clyde to the Smithsonian's National Museum of Natural History. He gave his bones to the museum so he could keep on teaching even after he died. I'll admit that I teared up a little bit reading about him.

Now, here's where it starts to get really interesting: The man's name was Grover Krantz. His last name is the same as my last name. The journalist in me got really curious: Who was this guy? What kind of person donates their bones to a museum? And could we be related? I looked him up online and found a big newspaper article about him in the *Washington Post*. It talked about how Grover had been an

important person in his field of anthropology and had written lots of articles and papers that helped us better understand how humans had evolved. It mentioned that he'd had three Irish wolfhounds— Clyde, whom I'd seen, and two more named Icky and Yahoo. And then there was one paragraph that changed everything. It said that he was one of the few experts in the world on Sasquatch. In fact, he had spent many years of his life in the woods with a spotlight and a gun, looking for one.

Whoa—*this guy's even more strange than I thought!* I said to myself.

I mean, this guy was a scientist! And a professor! But he believed in Bigfoot? Now I really had to know if we were related. Turns out, he'd been born in Salt Lake City, Utah, which was where my dad's family came from.

So I asked my dad, who wasn't sure. But he asked my grandfather, and guess what? Grover was his cousin! He remembered Grover showing up at the family picnics and measuring people's heads to see if he could learn something about evolution. He said Grover was always interested in human anatomy—our bodies and how they're put together—and my grandfather even remembered sneaking the hand of a skeleton out of his medical school classroom to give to Grover to study.

Oh my gosh—that meant that the skeleton with a dog, the professor who spent his life looking for Sasquatch, was my first cousin twice removed. Or was I his second cousin once removed? Whatever, the only thing that mattered was that we were related. That Sasquatch-seeking weirdo was part of my family! I had so many questions!

Lucky for me, not only did Grover leave his bones to the museum, but he also left all his papers and drawings, the letters he had received, a handful of yellowed newspaper clippings, and some photographs. Even better? The museum not only had all of Grover's Bigfoot papers but also things like his plaster Bigfoot footprints and his Styrofoam models of what a Bigfoot skull might look like. For a journalist like me, this was like finding buried treasure! But as I scanned over his notes and read some of the articles, I realized how little I knew about any of this. Because remember, up until just now, I'd thought of Bigfoot as a Big Joke. I mean, if you had asked me "What is a Bigfoot?" I'm not sure I'd even have known how to answer that question.

But Grover did. In an interview, someone asked him that very question, and here's exactly what he said: "Bigfoot is a large, massive, bipedal, hairy, higher primate. You could describe it as a gigantic man covered in hair and being rather stupid, or an oversized, upright-walking gorilla."

Yeah, you probably noticed that there are a lot of anthropology professor words in that sentence. Bipedal means it walks on two legs. Primate is a group of mammals that includes you, me, and all humans, as well as apes—chimpanzees, gorillas, orangutans—and lemurs, or just about anything that looks something like a monkey. So Grover was saying that Bigfoot is a big, hairy, apelike creature that walks on two legs. Which was kind of what I'd pictured.

Actually, once I started doing more research, I realized that that's how *everyone* pictured Bigfoot. Huge, with giant feet. Covered in hair—black, brown, gray, white. Sometimes with scary sharp teeth and snarling. Sometimes as a gentle giant. Most of the depictions seemed related to one of the most well-known pictures of Bigfoot of all time, which was taken by two cowboys back in 1967. In the photo, the Bigfoot is looking back over its shoulder at the two men, and that image became super famous. You've probably already seen it, but it looks something like this:

That Bigfoot is so famous, it has a name: Patty. Based on what we can tell from the photo, she's a lady Bigfoot. Which was something else I'd never thought about—that Bigfoot might be female. It made perfect sense, too, because if people have been seeing Bigfoot for centuries, in all kinds of different places, it couldn't be the same Bigfoot—there couldn't just be one! There'd have to be lots! Male Bigfoots and female Bigfoots and baby Bigfoots (or is it Bigfeet?)! How many Sasquatches are actually out there? Well, one anthropologist I talked to said there might be as many as two thousand wandering around the United States and Canada. How did he come up with that number? He told me he researched how much food a black bear needs and how many black bears can live in a certain area. Bigfoot and a black bear might be a similar size and have a similar diet, so he used the information about black bear numbers to make his best guess about Bigfoot numbers (assuming, of course, that Bigfoot would be a much rarer animal). Grover himself thought that there would be about one hundred bears for every Bigfoot!

Oh man, I have **A LOT** *to learn*, I thought.

I started by turning to the internet. Google "Bigfoot" and you're going to spend the rest of your life trying to read everything out there. And of course, not everything that's online is 100 percent true. Like the website that told me Bigfoot prefers to eat doughnuts. I mean, they're both delicious and portable, and I bet a lot of animals (not just Bigfoot) like them. I'm sure if you brought some camping with you, a Bigfoot (or some other animal) might steal them out of your tent. But doughnuts aren't something you can just find in the woods, and I can't imagine Sasquatch shopping at my local grocery store. There are probably lots of other things that Bigfoot can actually get its hairy hands on much more easily. So I knew I'd have to be careful about what kind of information I got from the internet. Did the information come from respected scientists or other experts? Did it make sense? Could I find more than one source that said the same thing? Even if the answer to all those questions was yes, it didn't guarantee that the information was accurate, but it made it more likely.

I wished Grover was still alive so I could ask him all my questions. Because not only was Grover a Bigfoot expert, but he was also one of the most important Bigfoot experts in the history of Bigfoot expertise! Even now, years after he died, he's still a legend in the Bigfoot community. People read the books he wrote and use his writings to argue in favor of Bigfoot's existence. He tried to bring science and serious research to a topic that a lot of people think is silly.

In fact, even Grover felt that way once. For his work, he studied what the marks left on bones might say about a creature's life and the environment it inhabited. He studied human anatomy (including measuring people's heads!) and could figure out how something walked or moved based on how their bones lined up, or what kind of imprint their feet left behind. But at that time, he didn't believe that Bigfoot was real. He thought the people who devoted their lives to searching the woods for an unknown primate were wasting their time. But while he was teaching at Washington State University, someone found a set of giant footprints outside an old mining town in Washington. They made plaster copies of both footprints—called casts—and brought them to Grover.

The casts fascinated him. While one of the feet looked like you'd expect (like a human foot, except HUGE), the other one looked like it had been badly injured. And, Grover wondered, if you wanted to make fake Bigfoot feet, why would you make one that seemed damaged? It was the sort of detail that would make a counterfeiter's job really hard—they would have to make a believable fake, AND a fake that showed a convincing injury.

"If someone faked that print, they'd have to put all these subtle hints of the anatomy design into it," he said. "They'd have to be a real genius." They'd have to be someone who was even more of an expert at anatomy than Grover himself. And after examining the footprints and puzzling over them, he started to think that Bigfoot might actually be out there, in the woods, leaving giant footprints behind. So in addition to his work on human evolution, Grover also started pursuing Bigfoot, scouring the forests and following up on tips that people around the country sent in to him. While the evidence seemed pretty convincing to him, he knew he would need something a lot more concrete before the rest of the world would accept it.

As more and more Sasquatch researchers heard about Grover's interest in Bigfoot, they started sending him things they thought were important clues, like plaster casts of feet, unusual hair they found stuck in bushes and tree bark, and pictures of bizarre structures deep in the woods. He analyzed them all, usually dismissing the samples as fakes or cases of mistaken identity. He examined footprints under magnifying glasses, to look for details that would be hard to fake, like dermal ridges (those lines on your fingers and, yes, even your toes—take your shoes off and look!). With the information he got from Sasquatch researchers, and using his own knowledge of human and primate evolution, he built models of what a Bigfoot skeleton might look like.

He also collected stories from people who said they'd seen a Bigfoot. He and Diane, his wife, would drive around the country to hear what Sasquatch-spotters had to say. I met Diane one afternoon at her home, and she remembered one story very vividly.

"One man we met was a backwoods farm-boy type. Just a humongous guy, probably 250 pounds, all muscle. He said he was out hunting and then he saw one—a Bigfoot," she said. He told her he'd never seen anything so big. "And this big tough guy looks right at me and blushes as he said to me, 'I pooped my pants when I saw it.'"

Diane laughed about the interview, but it reminded her how scary actually seeing a Bigfoot in person might be. And it turned out that a lot of people had those kinds of encounters. She said Grover got letters in the mail almost every day from people around the country who were absolutely sure they'd seen a Sasquatch, and they wanted to share their experiences with someone, anyone, who might take them seriously. People in their nineties wrote about something that had happened to them when they were just kids. They said they'd never told anybody before.

I didn't really understand why they'd keep a story like that secret for so long. My guess was that they didn't want people to laugh at them. Or maybe what they'd seen had been so terrifying that they just didn't want to think about it again. But over and over, Grover heard about how these people had seen something out of the ordinary, something unlike anything they'd ever seen before. It stuck with them, and even if they hadn't believed in Bigfoot before, they sure did now.

However, while these stories interested Grover, they weren't enough. Grover wanted evidence; the kind of evidence that would make other scientists take Bigfoot as seriously as he did. As a scientist himself, he knew he had to have a high standard. So he applied the same scientific techniques and thinking that he used in his regular work on human evolution to investigate the modern mystery of Sasquatch. Even though he thought Bigfoot was out there, he knew that that wasn't enough, and he also knew that sometimes the things that we really want to be real just aren't. What he needed was irrefutable proof.

The key to a scientific approach is finding evidence that *everyone* agrees is real. A footprint in the forest might be a clue to a creature's existence, but it might also point to other possible conclusions. Could it actually be a bear print? Or a human footprint that expanded in the mud? Or maybe it's from some trickster who's planting fake Bigfoot footprints? Even with something as mysterious as Sasquatch, the rules of science and facts still apply! Those rules are part of something called the **SCIENTIFIC METHOD**, which is kind of like the directions that scientists use when trying to explain something they don't understand. There are six basic steps:

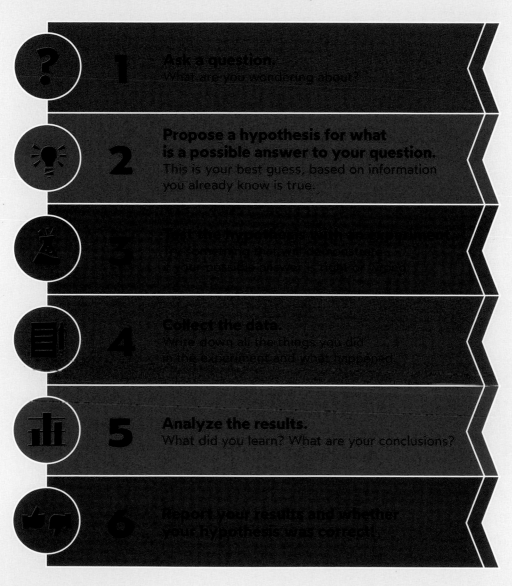

1 Ask a question.
What are you wondering about?

2 Propose a hypothesis for what is a possible answer to your question.
This is your best guess, based on information you already know is true.

3 Test the hypothesis with an experiment.
Try something out and demonstrate whether your possible answer is right or wrong.

4 Collect the data.
Write down all the things you did in the experiment and what happened.

5 Analyze the results.
What did you learn? What are your conclusions?

6 Report your results and whether your hypothesis was correct.

And if your hypothesis wasn't correct, you propose a different one and go through the steps again and again—this is called iteration.

It sounds like a lot of work, but a good scientist would never settle for less.

Neither would a good journalist. So as I explored the world of Bigfoot, I also had to think about those rules of science. I had to be very careful about verifying the facts, and making sure the information was true, before sharing it. I needed to remain objective—I had to look at the facts without letting my personal feelings get in the way. Meanwhile, the questions just kept piling up. If Bigfoot *did* exist, then why was it so hard to find? Why hadn't we ever caught one in a trap, or taken a good, clear photo, or found a dead one somewhere in the woods? What would happen if we did catch one? Should we put it in a zoo? And if Bigfoot *wasn't* real, why had the myth stuck around for so long? Why did people still go out searching?

There are lots of people, just like Grover, who've spent a huge chunk of their lives looking for what is probably the world's most elusive creature (only the Loch Ness Monster, in Scotland, might be more slippery). I was about to try the same thing. I didn't think I'd actually find Bigfoot—heck, I still wasn't sure I even believed in Bigfoot. But now, because of Grover, I considered myself Bigfoot-curious.

That weekend trip to the natural history museum changed everything. An exhibit on anthropology led me to Grover's bones, which led me to learning about Grover, which led me to Bigfoot. If you wanted to divide my life into two parts, I guess you could say there was Before Grover (BG) and After Grover (AG). And my life AG was going to be full of Bigfoot, starting with a look at Grover's specialty: how Bigfoot fits into the story of evolution. If, like Grover said, it's a big, hairy, bipedal primate, wandering around the forest, it had to evolve from something. Which means, at some point waaaaay back in time, humans and Bigfoot might have a common ancestor. It looked like I might have another long-lost cousin—and it might be your cousin, too.

WE ARE FAMILY

You are part of a **HUGE** family. Not just your parents and your brothers and sisters and your grandparents. Not even the entire human species, which we refer to as *Homo sapiens*. I'm talking about all the distant relatives we've had over hundreds of thousands of years—and there have been lots of them. In fact, if you go far back enough, you are related to every single living thing on this planet.

In 1859, a man named Charles Darwin came up with the theory of evolution, which explains how different types of plants and animals change over time, and even become new species. A species is a group of organisms, or living beings, that are closely related to one another; they share similar characteristics, and they can reproduce with each other.

Every time an animal has offspring, or a new plant springs up from a seed, or any sort of living thing reproduces, the new generation inherits traits from its parents. Traits are characteristics, like hair or eye color. Sometimes those new traits are very helpful to a species and make it easier for them to survive. For instance, let's say you have lots of squirrels living in the woods and they all have gray fur. Then a mutation occurs—a mistake or a change in a living creature's genetic makeup. Because of that mutation, some of those squirrels are born with brown fur. Those lucky squirrels blend in with the trees a little bit better and are less likely to be eaten by hawks and weasels. Squirrels that don't get eaten are more likely to have their own babies, and there's a good chance those babies will also have brown fur and be more likely to survive. After lots of generations, you'll have more and more brown squirrels.

Now let's say that the brown squirrels stay where they are, while the gray squirrels move to a different forest, where they are safer. The brown squirrels and the gray squirrels no longer hang out together. The food is different. The weather is different. The predators who hunt them are different. All these things have an effect on the gray squirrels' traits, and over time, they become more and more different from the brown squirrels. And if this goes on long enough (like thousands and thousands of years), those gray squirrels might become an entirely new creature—a new species.

A new species, such as a new type of bacteria, might arise after just a few days. Other species might take millions of years. And humans

took about 6 million years to evolve from our apelike ancestors into modern *Homo sapiens*. That sure seems like a long time, but we think the earliest type of life on Earth got started close to 3.5 BILLION years ago. That's a lot of years for all kinds of new species to emerge, adapt, and sometimes die out. Think about how many different types of living things are on Earth now, and we know that there are lots of others that don't exist anymore—some that we know about, like dinosaurs and saber-toothed tigers, and others that we don't know of, because we've never found any evidence of them. We're talking millions and millions of plants, animals, and insects, among other things.

Darwin believed that all the life on Earth that we see today descends from a very, very distant ancestor called, unimaginatively, our "last universal common ancestor" (LUCA). We don't know what LUCA was, exactly, or what it looked like. Our best guess is that it was a single-celled organism that lived between three and four billion years ago. Whatever LUCA was, its first offspring weren't much different. But like those squirrels, LUCA eventually evolved into lots of unique species, which evolved into other species, which evolved again, over and over, generation after generation, until you get the incredible variety of creatures we see on planet Earth today.

Some of those creatures are more closely related than others. Bacteria have a lot in common. Animals share similar traits. Plants do, too. By grouping living things with similar traits together, we can get an idea of how closely related they are. This is called classification, and there are seven levels: kingdom, phylum, class, order, family, genus, and species.

WHAT IS A MNEMONIC?

A mnemonic is a trick for helping people to remember certain things. It's pronounced "nuh-MON-ic"—the first *m* is silent. I use a mnemonic to remember those seven levels of classification, like this: <u>K</u>ing <u>P</u>hilip <u>C</u>limbed <u>O</u>ver <u>F</u>ive <u>G</u>iraffes <u>S</u>ideways—<u>K</u>ingdom <u>P</u>hylum <u>C</u>lass <u>O</u>rder <u>F</u>amily <u>G</u>enus <u>S</u>pecies. You could also say <u>K</u>ing <u>P</u>hilip <u>C</u>ame <u>O</u>ver <u>F</u>or <u>G</u>reat <u>S</u>oup if that's easier to remember. Or you can make up your own!

Another mnemonic I use is for the planets: <u>M</u>y <u>V</u>ery <u>E</u>xcellent <u>M</u>other <u>J</u>ust <u>S</u>erved <u>U</u>s <u>N</u>ine <u>P</u>izzas—<u>M</u>ercury <u>V</u>enus <u>E</u>arth <u>M</u>ars <u>J</u>upiter <u>S</u>aturn <u>U</u>ranus <u>N</u>eptune <u>P</u>luto. I learned this back when Pluto was still a planet. I guess I need to come up with a new mnemonic now!

Mnemonics can also be used to help remember how to spell things. For instance, the word "arithmetic" can sometimes be tricky (I always want to write "arithMAtic"), but if you remember this sentence, "<u>A</u> <u>r</u>at <u>i</u>n <u>t</u>he <u>h</u>ouse <u>m</u>ay <u>e</u>at <u>t</u>he <u>i</u>ce <u>c</u>ream," you'll know how to spell it correctly.

A good way to think about this is like a bookstore. You've got kids' books in one section, adults' books in another—you can think of these like the kingdoms. Then, in each of those "kingdoms," there are more sections, like fiction and nonfiction. Inside each of those you

Here I was now, it seemed, on the point of meeting face to face that monster whose ferocity, strength, and cunning the natives had told me so much; an animal scarce known to the civilized world . . . We were armed to the teeth. My men were remarkably silent for they were going on an expedition of more than usual risk.

Then he saw a massive, shadowy shape dart through the underbrush. He brought his rifle up to his eye but held his fire, knowing that he would only have one shot. A few seconds later, the beast broke out of the screen of forest just a dozen yards from where he was standing.

Nearly six feet high with an immense body, huge chest, and great muscular arms, with fiercely-glaring large deep gray eyes, and a hellish expression of face, which seems like some nightmare vision: thus stood before us this king of the African forest.

What? Did you think he was talking about Bigfoot? **Gotcha!**

Paul Du Chaillu had tracked down a gorilla. Until the mid-1840s, people outside of Africa thought that gorillas were a complete myth (just like Bigfoot). Yes, there were tales of the enormous ape that dated back all the way to the time of the ancient Greeks, but no one in Europe had ever seen one. Anyone who ever brought up the idea of one was more or less shunned in polite society (just like believers of Bigfoot). Things only started to change when a European traveler returned to London with specimens of unusually humanlike bones that were bigger than anything in the known anthropological records. Of course, locals in Africa knew the gorilla was real, but until European scientists could see one for themselves, they weren't going to take the gorilla seriously. This is where Paul Du Chaillu came in.

To prove its existence, Du Chaillu needed a specimen—a sample—and unfortunately, it was easier to bring back a dead gorilla than a live one. And when that specimen arrived in Europe, people went bananas. They wanted to know all about this supposedly mythical creature. They crowded into museums to see gorilla displays and even watched gorilla-themed ballets. It's easy to imagine their surprise and excitement. A creature that everyone thought came right out of a fairy tale was suddenly very, very real—and to top it all off, it kind of looked like us. It was like finding an untamed distant cousin of sorts.

Remember what I stated earlier: Even if we don't have acceptable scientific proof of something, that doesn't mean it doesn't exist.

If people previously thought the gorilla was a myth, despite all the stories, then what does that say about Bigfoot? What was the gorilla before someone found its bones? Just a legend. It was a joke. People would laugh at you if you took it seriously. That all changed when someone brought back a specimen. For people who believe in Bigfoot, the gorilla story is a beacon of hope.

After learning all about evolution, our distant relatives, and the fossil record, I started to wonder if something like Bigfoot might have existed once, even if it's not around anymore. But like those scientists in Europe who wanted a gorilla specimen, I also wanted to see proof, fossilized or furry, before saying Bigfoot is real. And that raised an important question: What sorts of evidence count, and what kinds of evidence don't?

THE
EVIDENCE

If you talk to any Sasquatch searcher, or squatcher (as they call themselves), many of them will tell you that the evidence for Bigfoot is overwhelming. "It exists. There's no question about that," said John Kirk, a Canadian policeman and an avid Bigfoot researcher. I met him at a Bigfoot conference in Willow Creek, California. John waved his hands wildly in the air as he talked. "If you're looking at evidence, what does that tell us about this animal? That they exist as much as Irishmen exist, that pizza is magnificent. These are all facts!"

Based on what I had just learned about evolution, I had started to think that Bigfoot could possibly exist, but I definitely wasn't as sure about it as John. I wanted to know more about all this evidence—what physical signs did we have that pointed to Bigfoot being real? And were they enough?

The most common bits of evidence of a giant, seven-hundred-pound primate in the woods have been the thousands of mysterious footprints found in streambeds and snowdrifts, around construction sites on logging roads, and deep in the woods, where there is simply nothing else around. People have reported finding tracks that are as long as eighteen inches, which would make Bigfoot's shoe size about a 37!

So what makes people think that they're Bigfoot footprints? Well, first off, they're really big. John said that the first pair he ever found were fourteen and a half inches long—that's more than a foot (ha)! Another reason is because the length between footprints—the stride—is really long. Where an adult human has a stride length of about thirty-six inches, or three feet, a Bigfoot might have a stride length of sixty inches, or five feet.

There's a lot of other information that people said they can get from footprints. For example, you might be able to tell how heavy a creature is, depending on how deep the print is. If it's a really clear one, you can get details about how the bones fit together, where the foot bends, or if it had been injured. Remember, it was a plaster cast of what looked like an injured foot that convinced Grover. He thought it would be way too difficult to fake the print of an injured foot in a way that would be convincing, especially to a scientist like him.

I kind of wondered about those plaster casts—how do you make them? And why? A couple of Bigfoot experts I met on a camping trip (more on that later) explained to me that they're an important part of how Sasquatch seekers collect footprint evidence. Footprints, as you may have already figured out, usually don't last very long. Rain will wash them away, and other creatures or people walking around will destroy them. So if you find one and you want to preserve it, you have to move pretty fast.

The first thing you need is a casting material, like plaster. You mix the plaster dust with water to make kind of a goopy sludge. Think of cement—if you've seen it poured out of a cement truck or watched someone mix it in a bucket, you know exactly what I'm talking about. Then you very carefully pour your goopy sludge into the footprint you found and wait for it to harden. Once it's all dry, you can pull the dried plaster out of the footprint. Now you have a permanent cast, and if you did it right, it will preserve the evidence and all the detail of the footprint.

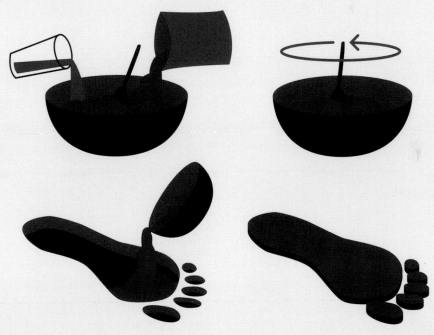

I'd seen Grover's plaster casts of Bigfoot footprints, in his collection at the Smithsonian. A lot of other people also have casts. They even trade them—they're kind of like the baseball cards of the Bigfoot world. Some of them are famous, too, like the plaster casts of the injured Bigfoot feet.

Because people have found so many footprints over the years, Grover thought there were only two good explanations: Either there was a real Bigfoot out there, leaving its real, big footprints, or there was a very clever and sneaky hoaxer, someone who understood both anatomy and how to make really good fake footprints (and has never been caught!). "The way I like to put it is this: The idea of a Sasquatch is ridiculous. But the alternative of a hoaxer is impossible. Therefore, the ridiculous must be true," said Grover.

But there are people out there who *do* make fake footprints. One of the most famous examples of this happened in California in the 1950s. A man named Ray Wallace strapped on a pair of boots that had wooden Bigfoot feet nailed to the bottom of them and walked around a logging camp. He set off a Bigfoot craze, as people came from all over to see if they could find the beast that left those prints. Wallace kept his joke a secret until he died in 2003, when his children confessed the truth. For squatchers, it came as a big blow because it made people doubt all the other footprint findings and Bigfoot sightings.

Grover also owned a pair of those wooden Bigfoot feet. He wasn't trying to trick anyone, but he wanted to show people how fake Bigfoot footprints would look. A wooden foot is stiff—it doesn't bend like a real foot. It also doesn't have those little lines, the dermal ridges, in the toes. And the weight wouldn't be right, either; a real Bigfoot footprint would be much deeper because the creature is so much heavier than a person. After Grover made his own fakes, he thought it would be pretty easy to spot someone else's.

But here's another question to think about: Assuming all those footprints out there were real footprints, what did they prove? Well, scientists would probably say that unless you see what made that footprint, or you have video or a photo of it, you don't know for sure that Sasquatch made those footsteps. A lot of people who are looking for Bigfoot get excited and jump to conclusions when they see footprints, but by themselves, footprints don't count as evidence. Grover knew this, which is why he kept searching for Bigfoot, even though he had a set of footprints that he definitely thought were real. He knew they were just one piece of the puzzle. So, I wondered, what other puzzle pieces were out there?

Well, another one that I heard about over and over were strange sounds. Grunts and whistles and snarls, big scary howls—a lot of the squatchers said they've heard some really odd sounds out in the woods, noises that didn't sound like any other animal they'd heard before. One of the most famous Bigfoot hunters out there, a man named Bob Gimlin, told me he heard sounds that would lift the hair right off the back of your neck.

Have you been camping? Or been outside at night in a park or a forest? I bet you heard some weird sounds, too. I definitely have—from creatures like coyotes and owls, bears and loons. I once even heard a screaming bunny, which might be the scariest sound I'd ever heard.

But the problem with strange sounds is the same problem we have with footprints: They're what we call circumstantial evidence—evidence that might prove Bigfoot's existence, but might also prove that something else was going on, like a hoax or a bear. If you don't see what's making the noise, or leaving the footprint, you can't say for sure what it is. So sounds weren't the kind of evidence that scientists could use.

CIRCUMSTANTIAL EVIDENCE: A TALE OF MICE AND MINTS

At Christmastime one year, my mom put out a big bowl of mints. Not hard, wrapped peppermints but these soft mints that came in a bunch of colors like pink, green, and yellow. They were shaped kind of like chocolate chips, and they had tiny little round white sprinkles on the bottom of them. They were pretty good, and I ate a few now and then. But they seemed to be disappearing pretty fast, and one day, almost all of them were gone.

My mom blamed me. She'd seen me sitting near the mints. She'd heard me say they were delicious. She'd even seen me eating them. And now the mints were gone. She thought that all this circumstantial evidence pointed right at me. I sure seemed guilty, but I definitely hadn't eaten all those mints!

> We looked closer, and we saw a trail of those round white sprinkles leading under the couch. We lifted up the couch cushion and saw a pile of mints with little teeth marks in them. We looked around the mint bowl and found some little brown mouse poo-poos (*ewwww*). And now the circumstantial evidence was pointing at a mouse.
>
> But we didn't have direct evidence until we set a mousetrap near the mint bowl—and caught one. Which shows you that circumstantial evidence might give you some ideas about what's happening, but you can't know for sure until you have direct evidence.

Luckily, people do see things, or at least they think they do. And they even try to take pictures. There are videos and photos all over the internet that supposedly show Bigfoot walking through the woods, or lurking behind a tree, or running away. Unfortunately, they all have one thing in common: They are VERY blurry.

I asked some Bigfoot researchers about this, and they told me that these kinds of blurry shapes are called "blobsquatches." They're pictures of something that somebody said was a Bigfoot, but you couldn't quite tell for sure. It could have been a stump. Or a shadow. Or a bear. Or a guy in a gorilla suit. Or just nothing at all. But isn't that a little odd? That everyone is so bad at photography? I mean, these days, a good chunk of the population is walking around with a smartphone in their pocket, so they have a camera at hand all the time. So why isn't there a single, crystal-clear image of Bigfoot out there?

I asked Dr. Jeff Meldrum this question. He's also an anthropologist and a professor at Idaho State University, and just like Grover, he thinks there's a very good possibility that Bigfoot is real. In fact, he talked about this with Grover and had a chance to learn from him. Dr. Meldrum said the reason all the photos are so bad is because most of the people who see one are completely caught by surprise. They're so excited and nervous about what they're seeing that they just can't take a good picture. It's also very hard to get good photos of rare wildlife, especially wildlife that doesn't want to be seen. Which is true! I once knew a photographer who had been told to take pictures of wolves in Yellowstone National Park. He spent weeks chasing after them but could never get close enough to get a good picture. Eventually, he had to go to a wildlife park—almost like a zoo—to get his photos. There's no Bigfoot zoo (that I know of) where someone can go to take better photos of Bigfoot!

These were all good points, but I still wasn't sure. People have been out looking for Bigfoot for decades, cameras in hand. Plus, now there are all these drone cameras and wildlife cameras and security cameras. Those are machines—there aren't any people behind the lens, taking those photos—and machines don't get nervous. And like I said, so many people have phones with cameras right in their pockets. Surely someone can get a good photo? I started to think that maybe the problem isn't the cameras. Maybe Bigfoot is just blurry in real life!

I wasn't so impressed with the photographic evidence. But there was one video that really puzzled me. Only a minute long, it was shot more than fifty years ago, in the forests of Northern California. It's called the Patterson-Gimlin film, and it was named for Roger Patterson and Bob Gimlin (that guy who said he'd heard spooky noises, too). They were two rodeo cowboys who took a trip to California to look for Bigfoot, way back in 1967. I sat down with Bob Gimlin, to see if he would tell me his story in person. It's been decades since it happened, but he remembered every little detail about that day, October 20.

"Roger was in front of me on his horse. And it went bananas. That's when we saw a giant, hairy thing walking in front of me at one hundred yards, and it just kept walking away. All I'm thinking is that these things really do exist! And it's walking away from me like a human being," he said.

Gimlin remembered jumping off his horse and grabbing his rifle. He didn't want to shoot the creature, but he also didn't know what it was and wanted to make sure he could protect himself if it attacked. Meanwhile, his friend Roger Patterson pulled out the camera he had in his saddlebag and started filming the creature as he ran after it. Because of that, most of the video is very, very shaky as Roger goes bounding through the clearing. Then he stopped, and you can see this big, hairy beast lumber off into the woods.

The film made history. Just about everyone who grew up in the 1970s saw it when they went to the movies. And one frame in particular—a frame is like a single photograph pulled from a movie—known as Frame 352, shows the creature looking over its shoulder, with one arm in front and one arm in back. Remember that famous photo I mentioned earlier in the book? This is Patty, and she's named for Roger Patterson. And even if you've never seen this film, you've probably seen Patty—her image is on bumper stickers and socks and pizza. It's **THE** image of Bigfoot.

Even though the film became famous, there was a lot of debate about it. Some people thought it was a hoax and that Roger Patterson and Bob Gimlin completely made it up. They said that it was just a man in a gorilla suit. Other people believed Bob and said this film was—and still is—the best piece of Bigfoot evidence out there. Either way, Bob has stuck to his story for more than fifty years, and what he said convinced me that he definitely saw something weird, even if it wasn't Bigfoot.

One of the people who believed the film was real was Grover. He studied every detail about that movie, going frame by frame to take measurements. In his opinion, it looked pretty authentic.

"The walk is peculiar," he said. "It leans forward and keeps its knees more bent than is normal for a human. It lifts its heel higher when it lifts its foot off the ground than a human does. But it swings its arms in a human manner." Grover used the film to estimate Patty's weight (five hundred pounds) and her height (six feet, five inches tall). To prove that it couldn't be a hoax, he tried to imitate the walk for a television news crew, and it looked very, very silly—not at all like Patty walking through the California forest.

The Patterson-Gimlin film was the last time we had anything even close to a clear image of Bigfoot, but that was decades ago! And while I was certainly intrigued by the movie, I didn't think it was enough to say that Bigfoot definitely exists. Neither were those footprints and sounds and photos. They were all circumstantial evidence.

What we need is direct evidence, and most scientists would agree. If you're going to declare a new species, you have to have a flesh-and-blood example of it—like the gorilla that Paul Du Chaillu brought back from Africa. This means we need an actual Bigfoot—what scientists call a type specimen. Grover knew this, and while I know it sounds sort of horrible, he was famous for saying that the first person to see a Bigfoot should kill it. (No wonder Bigfoot is hiding from us!)

"Almost any scientist will tell you, in no uncertain terms, that you will only prove the existence of Sasquatch by bringing in a body, or at least a substantial piece of one," he said. "No other evidence is proof."

A body. Or at least a big piece of one. That's the only thing that would stop the debate about Bigfoot's existence—every other type of evidence out there can be misinterpreted, or faked. But shooting Bigfoot is a pretty unpopular idea for a lot of people. In fact, when Grover started saying that someone should kill a Bigfoot, he got a lot of very angry letters. Never mind whether they believed in Bigfoot or not; they were furious about the very idea of shooting one. One man wrote, "Someone might want 'final proof' of a Grover Krantz. Do you suggest that someone shoot you? Leave Sasquatch alone. If the creature does exist, learn to exist with him (or her)."

Still, the squatchers I spoke with knew that if they wanted their search to be taken seriously, they would have to provide some sort of type specimen. They talked constantly about how they would capture a Bigfoot, dead or alive. Kathy Strain, an anthropologist I spoke with, said she's pro-kill because she's a scientist and that's just how science is done.

Another lifelong Bigfoot hunter, named Peter Byrne, who used to hunt tigers in India, doesn't want to kill one. "I don't shoot anymore," he said. "If I saw one, I wouldn't shoot. Suppose the one you shoot is the very last one there is?" That would be a very big problem (and more than a little sad). Plus, if you shoot it, it can't lead you to others of its kind and you wouldn't be able to see it interact with its habitat.

In an ideal world, Grover hoped that he wouldn't actually have to pull the trigger. He hoped that someone might accidentally stumble across a Bigfoot that had died of natural causes. But even Grover knew this might not happen. More and more humans are exploring the forests—hiking and hunting and camping—and no one has stumbled across a Bigfoot body yet. Squatchers have a good argument for why that might be—scavengers.

Peter Byrne, who has been looking for Bigfoot and its Himalayan cousin, the Yeti, since the 1950s, told me that when something dies out in the woods, it gets eaten by scavengers. "The principal garbage man is the bear. After that we have coyotes, porcupines, crows, and ravens, just to name a few. So when a Bigfoot dies, you have four hundred pounds of remains lying there, and it doesn't take long to get disposed of."

The yummy smell of a dead Bigfoot body would draw these other animals in, and they'd gobble it up like an all-you-can-eat buffet, which explains why we haven't found any bodies. He made a good point—I've gone hiking a lot and I've almost never seen the body of a dead animal. I've mostly only seen a few bones, and usually not enough to tell what kind of animal it was. They don't hang around for very long!

After everything I'd heard so far, I didn't feel very hopeful about find-ing real proof of Bigfoot's existence. But something has changed since Grover was searching in the woods for Bigfoot. There's something new that could answer the question about the big guy's existence, even if we don't have a body. It's another type of physical evidence that you can't see with your naked eye. You've probably heard of it before: It's called DNA. And remember those nests? If Bigfoot or a family of Bigfoot (Bigfeet?) made them, they should be covered in Bigfoot DNA. I hoped I'd find out!

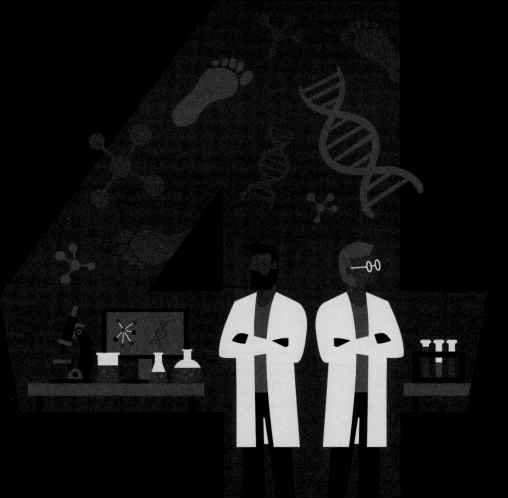

BLUEPRINTS, NOT FOOTPRINTS

Remember at the beginning of the book, when Shane showed me where scientists had cut a big, pizza-shaped wedge out of one of those giant nests in the forest? I didn't quite understand why they'd taken it or how they'd get any information from it—wouldn't it have been better to leave the nests intact, all in one piece?

Well, it turns out there's a lot to learn from taking a sample like that, as two scientists explained to me. One of those scientists was Dr. Jeff Meldrum, the professor who explained why the Bigfoot photos all turn out blurry. "I have to say I'm not 100 percent," he told me. "But I am quite convinced—I'm convinced up there to the 99.99 percent."

The other scientist was Dr. Todd Disotell, who's also an anthropologist, but he's pretty sure that Bigfoot is *not* real. "The chance that Bigfoot is out there is probably not zero, but it's close to it," he said. "We have zero biological evidence."

You might have noticed that neither of these scientists said that Bigfoot is definitely real or definitely imaginary, and that's because, like we learned in the last chapter, we need the right kind of evidence. "We don't have definitive proof and that means having a type specimen," said Dr. Meldrum. "So until there's physical evidence that's convincing, we can't say for sure."

Dr. Meldrum hoped that the nests might provide that piece of evidence. A few months after the nests were discovered, Dr. Meldrum traveled out to the Olympic Peninsula to see them. His experience was similar to mine. "They were in a cluster, on this ridge that overlooks about a twenty-to-twenty-five-foot drop down to a creek," he remembered. "They were hidden behind this huge, tall screen of vegetation." He said there was a total of about twenty-one nests—some were big enough for a person to lie down in, like the first one Shane showed me, and others were much, much smaller and tucked into low branches, just off the ground. He described the nests as being very complex. "The understructure [which is like the skeleton of the nest] was made of branches of all sizes, from twigs up to sticks an inch across. Then other materials like leaves had been piled in," he said. "And then the branches of huckleberry bushes were braided in around the edges."

Because there were so many nests, Dr. Meldrum felt comfortable taking a sample out of one of them. He used a set of pruning shears—like a giant pair of scissors—to cut out a wedge in the nest. Then he and his assistants wrestled it into a plastic bag, which they took back to his lab, where he and his team looked through it very carefully, searching for hairs, bones, bugs—anything that might be a clue about what had built the nests. He did say they'd found some strange hairs, but hair can be tricky to identify. Think of all the different kinds of hair you have on your body—on top of your head or your eyebrows or inside your nose or on your arms. Those hairs are very different from one another. Animals are the same, and it can be hard to tell what hair comes from what animal.

Dr. Meldrum also took a few other pieces of the large sample that he sealed into plastic containers with screw-top lids. These samples were mailed off to be examined for something very special—**DNA**.

Back in 2008, a group of paleontologists—scientists who study the remains of ancient plants and animals—were exploring a cave in the frigid north of Siberia. They were looking for evidence of ancient dogs, called canids. As they carefully combed through dirt and rock on the cave floor, searching for clues, they uncovered a single, teeny-tiny finger bone, less than one inch long. That might not seem all that exciting—I mean, it's not a pile of gold or some ancient cave paintings. But dogs don't have finger bones, and the scientists wondered what, exactly, they'd found, so they sent it off to a laboratory. And when those paleontologists received the results, they got a real surprise.

That tiny little bone? It didn't belong to anything else we knew about. It came from a completely unknown human relative—one that we've since named Denisovans, for the cave where scientists found that finger bone. How did these scientists figure out that this was the bone of an unknown species? I mean, all they had was part of a finger! Welcome to the exciting world of DNA.

You've probably heard of DNA before. It gets mentioned in detective stories and crime shows. In the *Jurassic Park* movies, paleontologists used DNA to bring back long-extinct dinosaurs. Maybe you or your parents have had your DNA tested to find out if your relatives came from Italy or China or Ecuador. DNA is one of the fundamental building blocks of life on this planet. Every cell, in every living creature on Earth, has DNA in it. Think of it as a sort of blueprint for life—the instructions for building a person, wombat, cactus, jellyfish, or, yes, even a Bigfoot (*if* it exists). But what *is* it?

WHAT DOES DNA LOOK LIKE?

Think of a ladder—two straight sides with the rungs between them. Now twist that ladder into a spiral. This is what DNA looks like, and its shape is often referred to as a double helix. The two parallel sides of the ladder are what help DNA keep its shape, and they're made up of something called the sugar-phosphate backbone. The rungs of the ladder are made up of chemicals called bases, and there are four bases in DNA: adenine, thymine, guanine, and cytosine—A, T, G, and C. Each rung of the ladder is created by two bases joined together, called base pairs. A is always paired with T, and G is always

> paired with C. These base pairs get repeated over and over again, in a different order, and that order is like a code that tells our cells what to do.

Those three letters—DNA—are an acronym that stands for *deoxy-ribonucleic acid*. That didn't clear anything up for me, but what did help was thinking about books. Think back to our bookstore from Chapter 2—we used it to explain how different species get classified. You had the adults' section and the kids' section, which were like the kingdoms, all the way down to a very specific type of book, which is like a species. Now think of an individual book as one creature—maybe you. All the words, letters, sentences, and paragraphs make each book different from other books. So you can think of your DNA like a book—a book of blueprints for what makes you unique. Your "book" is organized into really long chapters that give instructions for different parts of the body. These chapters are called genes. One gene might tell the body how to construct an eyeball, while another one will have instructions for the skeleton. The language that this book of genes is written in uses chemical letters called bases. What's interesting is that, unlike a real book written with the twenty-six letters of our alphabet, the DNA book only has four letters: A, T, G, and C. Those letters get used over and over again, billions of times. It's how the four letters—those four bases—are combined that makes each book different, and the order of all those letters is what's called a sequence. A human book, like yours, has about 23,000 chapters (genes) with about 3 billion letters (bases). And every creature is different. The tiny water flea, which isn't much bigger than a grain of sand, has 31,000 genes but only 200 million bases.

Every cell in your body has a copy of this book of genes—that's how they each know what to do. So your unique DNA is in your blood, your bones, your skin cells, your poop, and your boogers, too. So, for example, let's say you're some kind of ancient human relative, living in a cave in Siberia, and when you die, your bones stay there. Then maybe one day, a paleontologist finds those bones and they use special and extremely powerful microscopes to look at the cells, and the DNA—the book—inside. Then, using super-fast computers, they can compare that book with other books that they already have copies of to see if they match. If they do, the scientists know exactly what kind of species they've found. And if they don't? Well, that means they *might* have found something new. Like a Denisovan.

This is why DNA is so exciting for everyone who's searching for Bigfoot. We learned that scientists need a type specimen—a body or a big piece of a body—to identify a new species. DNA could also do the job. So why haven't we used this before? Why didn't Grover use DNA?

Well, while we've known about DNA for a long time, it's only recently that we've been able to understand it. For one, it's really tiny. It's also really long—remember, a strand of human DNA is three billion letters—so it took us a while to read it. Scientists used to use chemicals

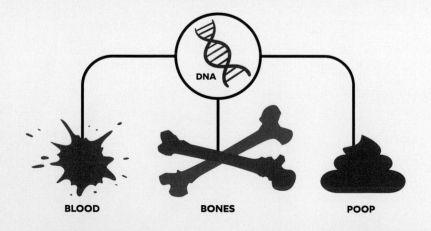

BLOOD   **BONES**   **POOP**

to extract DNA from cells, but DNA is a long, delicate string. It broke apart easily and made it hard to figure out. And on top of all that, it was really expensive to do this—it cost hundreds of thousands of dollars. As technology has improved over the years, we've gotten faster computer programs that can read DNA, which makes this process both cheaper and easier.

But decades ago, when Grover was doing his Bigfoot research, that kind of DNA technology didn't exist. He knew that it would happen eventually, and that someday DNA research might be really important to the kinds of work he did. In one interview, he said, "There are some new techniques where I'm assured they can extract the DNA from animal feces and identify everything that it ate."

Grover would have been astounded by how far the science has come. I really wished he was still alive now to see how we can do a lot more than just identify what's in poop. And he would be really excited to meet Dr. Todd Disotell. As a primatologist, Dr. Disotell studies primates—lemurs, monkeys, apes, and us humans. But rather than spending all his time in the jungle, watching monkeys, he studies them using DNA. And he has learned some really awesome stuff.

"The coolest thing to me was a study on the gunk between Neanderthals' teeth," he said.

Ewwwwww, I thought.

He explained that scientists in Europe had found some skulls with fossilized plaque stuck to their teeth. Even though the skulls were really old, that plaque still had DNA in it, because the skulls had been preserved in the right kind of environment—not too hot, not too wet. Also, teeth (and bones) are some of the best places to find DNA on ancient remains. The scientists in the country of Belgium scraped the tooth plaque off the skull they found, analyzed the bits of DNA in it, and discovered that those Neanderthals were eating woolly rhinos, deer, and all sorts of plants. But scientists who did the same thing to Neanderthal skulls in Spain found that those individuals were eating seafood and completely different plants.

I don't know about you, but this sort of science *really* made me want to brush my teeth. I couldn't stop thinking about all the strange DNA that might be in my gums.

But there wasn't any time to floss as Dr. Disotell explained why this was so amazing. Because the technology is so good, scientists like him no longer need an entire strand of DNA—all three billion letters—to figure out what kind of animal or plant they're looking at. They can figure it out with way fewer letters—as few as fifty! What this means is they can collect dirt from a cave floor, or scrape plaque off a tooth, and still be able to get enough DNA to figure out what kind of animals and plants have been there. This is called environmental DNA, or eDNA for short. Scientists don't even have to see the animal that lived in that cave—they just need samples from the area where the animal has been.

When your skin peels, or you get a bloody nose, or you poop in the woods, that DNA gets left behind. Or if a strand of your hair falls out at just the right moment and has the root still attached, it could also have some DNA on it. And if a scientist found those skin cells or blood or poop, even if they never saw you, they'd be able to use that DNA sample to identify that a *Homo sapiens* had been there.

It works everywhere. Want to know what fish are in a pond? Take a sample of the water and run it through the test, and you'll come up with little bits of fish DNA, along with the DNA of everything else that had been there. As long as we know the DNA—the book—of every species there, then we have the tools to identify each one.

Because everything on the planet is related, it means that most of our DNA is remarkably similar. If you lined up all the DNA from every mammal on Earth next to one another and compared them, you would only find a few, tiny differences. (This is why computers are helpful when we do this—it would take forever to do without them.) Human DNA is 99 percent the same as chimpanzee DNA. Other primates are a little more different—we might be only 94 percent the same. And you might be surprised to hear that more than half our DNA is exactly the same as a banana's, because at some point everything on the planet—including plants—came from LUCA, that Last Universal Common Ancestor that we all shared three or four billion years ago. Luckily, even a small difference in the order of those four letters that make up DNA—the

A, T, G, and C—can help Dr. Disotell tell the difference between the species in his samples.

You might be wondering about those nests that the Bigfoot researchers found—the ones I got to see. If a Bigfoot had been living in one of those nests, it would have left behind hair and skin cells and saliva (as it drooled on its Bigfoot pillow). This is why scientists took a big, pizza-pie wedge out of one of the nests—they can use eDNA to analyze what might have built it and lived in it. And guess who was asked to analyze that DNA? That's right—Dr. Disotell. Those Bigfoot researchers out on the Olympic Peninsula hired him to use his powerful microscopes and computers to see what he could find.

But, you might ask, how can we know we have Bigfoot's DNA if we don't actually have a Bigfoot sample to compare it to? We don't have any idea what Bigfoot DNA looks like—we've never read that book. This is definitely a problem, but it's one that scientists can deal with. As they look through all the DNA in a sample, they'll be able to eliminate the DNA that obviously belongs to other animals and plants. Eventually, if there's an unknown species lurking about in the forest, those scientists will be left with a bit of DNA that doesn't match anything we already know about. A new type of DNA *might* mean that we've found Sasquatch.

Now, that's a lot of DNA to go through. There are loads of bacteria living in that nest. Ninety-nine percent of everything they find is going to be bacteria, because there are more bacteria on Earth than anything else. But bacteria DNA is way different from lizard DNA or squirrel DNA or primate DNA. So Dr. Disotell programmed his computers to just look for certain kinds of DNA, which makes the process go a lot faster.

And in the case of the nests, Dr. Disotell thought that if they were made by Bigfoot, the best DNA to look for would be primate DNA. Although Dr. Disotell doubts Bigfoot exists, he's willing to consider the possibility. After all, we've found other creatures that we didn't think existed, so he wants to keep an open mind. He shared an example of a species of gorilla that he found in Africa. No one could get close enough to take a photo of these animals—all the photos were just blurs (like Bigfoot). But the researchers could take samples of poop from the gorillas' nests. And from studying that poop, they were eventually able to identify each of the individual gorillas in the group. All without getting close to an actual gorilla! Paul Du Chaillu would have been amazed.

"Poop is poop," said Dr. Disotell. "DNA is DNA. So if I ever found unknown DNA, I could tell you what branch of the evolutionary tree that it comes from," he explained. "It's exactly the same with Bigfoot."

WOW! I thought. *What if this is finally the piece of evidence that we need to prove Bigfoot exists?*

There had been so many times that Grover, and all the other squatchers, felt close to finding our mysterious friend in the forest. Maybe this would be the moment. But even if it *wasn't* Bigfoot, those nest samples could tell us something else totally interesting. Maybe we'd learn something new about the forest that we didn't know about before. Maybe we'd find a type of bear that builds fancy nests! Or giant squirrels that don't like living in trees!

EYEWITNESS

OK, everyone. Grab your flashlight and your blankets. Lock your doors and windows. And get ready for something spooky. Because the Bigfoot stories you're about to read will have you looking over your shoulder, wondering what's out there, going bump in the night. Ready?

Late one night, Hal Halderman and his family drove down a remote and winding highway in the mountains of California. Rocky peaks towered in the distance as the family van wound around curves on the mountain pass. In the dark, the Haldermans couldn't see much—just what the headlights lit up as they passed by, mostly the trunks of massive pine trees, some deer, and even a couple of bears. And then, as they came around a bend in the road, Hal caught sight of something standing under a tree, right on the edge of the road.

"It was taller than the van, well over seven feet tall," he told me. "It had a massive upper body, huge triceps hanging off its arms, and was very lightly colored, almost white, but also dirty and gray. It was reaching up into a tree." The headlights illuminated it for just a second, and then the creature sank back into the shadows. Hal slammed on the brakes and brought the van to a standstill. Then he turned to his wife.

"Did you see that?" he asked, his voice creaking just above a whisper.

"Yes!" she replied, terrified.

Hal wanted to get a second look, to make sure he'd seen what he thought he'd seen. He threw the van into reverse and started backing up. But his wife did not like this idea, not one bit. In a panic, she started screaming, yelling at him not to go back. They needed to get away. All that racket woke up their kids, who had been sleeping in the back seat.

"What was it, Dad? What was it?" they asked.

"I think we just saw a Bigfoot."

But Hal could tell his wife was scared, and he felt a little nervous, too. So he didn't go back. And he's been kicking himself ever since.

"I want to see it again," he said with a sigh.

It's been thirty years since that drive in California's mountains, and Hal thinks about it all the time. He had never believed in Bigfoot before, but he knows what he saw that night—and that moment changed everything for him. A few years after that California trip, Hal moved his family from Arizona up to Oregon—into Bigfoot territory, so he could go out and search as often as possible. He hoped he would see the creature again, just to prove that he wasn't crazy. When we met near his home in Oregon, he was still making regular Sasquatch-seeking trips out into the woods.

Hal is definitely not alone when it comes to seeing Sasquatch. A woman in Idaho said she drove off the road after Bigfoot ran across her path. Another woman in California said she saw several Bigfeet in the mountains and demanded that the government protect the species. A couple driving in Ohio caught sight of a tall, reddish-brown creature crossing the road behind them. And for some people, like Hal, that one sighting changed their whole life. Lots of the squatchers I met became squatchers because of just one sighting. They claim to have seen a Bigfoot once and now can't stop trying to see another one.

You know who heard a lot of these stories over the years? Grover. After I found out he was my cousin, and I started learning more about him, I went to talk to his wife, Diane Horton. You might remember her from near the beginning of this book. She said Grover used to get all these letters in the mail, from people who swore they'd seen a Bigfoot. She also used to go out with Grover to interview people who had reported a sighting. Grover would hear what they had to say, ask questions, and listen for any part of their story that could be useful to his research. He thought many of those sightings were probably a case of mistaken iden-tity—someone had seen a bear or a stump. Some of the stories were just too crazy for Grover to believe, like the idea that Bigfoot could speak English or magically transform into another animal or turn invisible.

But a handful of stories caught Grover's attention, mostly the ones that happened in the Pacific Northwest, where Grover lived. When I went through his papers at the Smithsonian, I found a list of supposed Bigfoot sightings that he thought might be the real thing and were worth more investigation. For each sighting, he wrote down the date, the location, and the name of the witness. Most of the names on the list had big *X*s through them—after doing more research, Grover didn't think they were real Sasquatch sightings. But a handful had question marks, or "*maybe*" written next to them in his blocky handwriting. These stories clearly left him scratching his head.

What did these people see? When I heard Hal's tale, I could tell that he absolutely, 100 percent saw *something*. He wasn't lying or making things up. In fact, all the people I talked to had some sort of experience that they couldn't explain any other way except with Bigfoot. And some of these people had a lot of experience in the woods, which meant that they knew a lot about what animals live there, what noises they make, and how they behave. They had looked for any other explanation, but they couldn't find one. That was what happened to John Mionczynski.

John worked as a wildlife biologist for the U.S. Forest Service out in Wyoming. He spent weeks backpacking through the wilderness, studying bears and bighorn sheep. He once even offered himself up as bear bait! John plays the banjo and lives in a tiny cabin that he built himself, where he still uses a wood-burning stove for heat and cooking.

He's a true mountain man, and he probably knows more about the natural world than anyone I'd ever met. That's why his story might be the one I believe most.

Back in the 1970s, when John was in his thirties, he was doing some research on bighorn sheep way out in the Wind River Range, a mountain range in Wyoming. He'd found a place to camp for the night and set up his sleeping bag and tent. Now, he'd borrowed the tent from the supply closet at the U.S. Forest Service office, where he worked, and the last guy who used it had dropped bacon on it. But John didn't have time to clean the tent off, so he knew he was going to be camping with a big, tasty bacon grease stain on his tent. Definitely a treat for bears—and a risk for him!

That night, John climbed into his sleeping bag and had just started to fall asleep when he heard some odd breathing right outside his tent. Of course, he immediately thought it was a bear, especially because of the bacon. The moon was full that night—it was very bright outside—and when the animal got closer, John could see its shadow on the tent wall. It certainly looked like a bear! And it was pressing into that bacon grease stain! Since John knew all about bears and how to deal with them, he decided to just scare it away. He let out a loud yelp and smacked the creature through the side of the tent. He hit something soft—maybe its nose?—and it ran off. But he could still hear it breathing, which meant it wasn't too far away.

About twenty minutes later, it came back and again pressed into the side of the tent. So John hit it again, and once more it ran off, but not as far this time. And then it came back a third time, and this time, the shadow was over the top of the tent and it looked like it was

walking on two legs. John thought it could be a bear hanging on to the branch of the pine tree that stuck out over the campsite. Bears, he knew, could grab branches and walk on two legs.

But then something pressed into the side of the tent again. John started to worry that maybe this was a mother bear with cubs, which can be very, very dangerous. But he smacked the side of the tent again anyway, and instead of hitting something soft, like the last two times, he hit something hard. As soon as he did, a shadow came down over the top of the tent, and it was a silhouette of a hand that was about twice the width of a human hand, with a thumb and very thick fingers.

And John knew that this was no bear.

Bears don't have hands. They definitely don't have thumbs, not like ours.

Uh-oh.

John scrambled to get out of his sleeping bag, to see what was outside his tent. His movement sent the creature running back into the nearby trees, so he couldn't see it. But there was no way he was going to climb back inside that tent again. He grabbed his sleeping bag and went to sit by the fire he'd made earlier. And then he waited and waited. He could still hear it breathing, but it wasn't getting any closer. He didn't think it was a bear anymore, but he didn't have a clue what it actually was.

"The thought of Bigfoot did not even enter my mind at that point," he told me.

Can you imagine what that would be like? All alone out there in the woods with some heavy-breathing, unknown creature lurking nearby? I'm sure I'd be really scared and nervous. There's no way I could ever fall asleep. Except maybe after a while, when things are quiet and the fire is warm and relaxing and I'm really tired after a long day of hiking . . . That's exactly what happened to John, and he started to doze off.

He wasn't sure how long he slept, but he woke with a jolt when something hit the ground next to him. He opened his eyes and by the light of the moon he saw what he thought was a pine cone fall out of a tree. Then, just seconds later, another one came flying out of the woods at him. Then another. And another. Over the next ten minutes, more than twenty pine cones landed in the fire and all around him. Something was throwing pine cones at him, and he definitely didn't think it was a bear. At this point, John didn't know what to do, so he just stayed still. He hunkered down with his sleeping bag over his head until eventually the creature shuffled off and left him alone.

As the sun came up, John looked all around the campsite to see if he could figure out what kind of animal had been harassing him all night. He didn't find any footprints or any other clues. But when he got back to his office a few days later, he told a few people about the mysterious attack, including his boss. And it was John's boss who put the idea of Bigfoot into John's head. He said he'd heard other recent stories about a giant, apelike creature running around that same area and scaring people. Maybe John had come across a Bigfoot.

John's story made the hair on my neck stand up. After hearing him tell it, I was almost ready to say that Bigfoot is real.

And there are so many other people—thousands!—who claim that they've seen a Bigfoot. Or smelled one (according to the witnesses I talked to, Bigfoot is apparently a very stinky creature). I found this map, which shows where people have reported recently seeing (and smelling) Bigfoot.

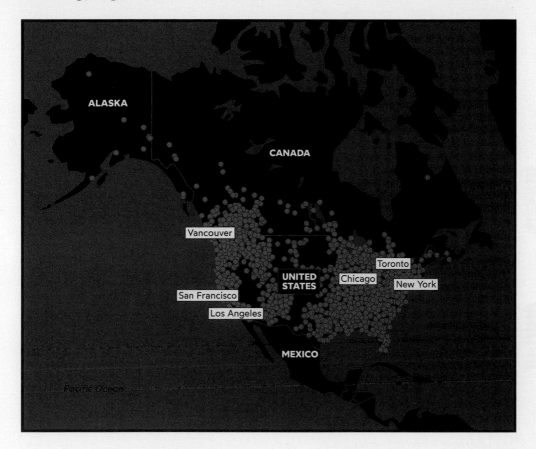

However, stories about Sasquatch go way back. Native Americans and First Nations peoples have been telling stories about a big, hairy, apelike thing in the woods for hundreds of years. Harvest Moon is a member of the Quinault Indian Nation, which is part of the Coast Salish group of Indigenous peoples. You might remember that we think the word *Sasquatch* is the Anglicized version of *Sasq'ets*, a Coast Salish word that means "wild man." Harvest Moon grew up with stories about Bigfoot, and she knows at least half a dozen people who have seen one, which can be a little scary.

"When you see a bear in the zoo, you're pretty comfortable," she said. "But when you're out in the woods and you see an animal face-to-face, it's quite a bit different. Seeing a Bigfoot can change you." She told me about the time her son saw a Bigfoot, crossing the road. It scared him so badly that he didn't go back into the woods again for a long time.

Tom Sewid is a member of the Kwakwaka'wakw—a First Nation's Indigenous people in Canada. They call Bigfoot Dzunuk̲wa (pronounced "TCHOO-nah-quah"), and Tom told me that when you go out in the woods, if you hear certain chirps, hollers, and whoops, you know it's Dzunuk̲wa. He said she's known as the wild woman of the woods, a female Sasquatch.

"Every family has a story or interpretation of an ancestor seeing that creature," he said.

Those stories get passed down over generations, but they aren't just from a long time ago—the Kwakwaka'wakw still claim to see Dzunuk̲wa to this day. Tom even had a story of his own.

A few years ago, he and his girlfriend were out fishing with some of their friends. They planned to spend the night on the boat and had dropped their anchor close to shore in a quiet bay. As they cooked up their catch on an old camp stove on deck, they started hearing

eerie whistles and chirps. On the beach, they saw two big shadows walking in the moonlight.

Tom remembers that the air smelled really bad, like low tide. It stank like rotten seaweed and murky water and dead fish. But it was high tide, and Tom realized that the horrible smell was coming from the Dzunu<u>k</u>wa. Tom and his pals flipped on the boat's spotlight, to see if they could get a better look. Two huge, hairy figures were picking through the clam beds, apparently looking for their evening meal. When the light came on, the creatures dropped to the ground and tried to hide. Tom wanted to get a photo, but his camera was out of film. He thought about shooting one, to bring back as proof, but he couldn't do it. In the end, all he brought back from that encounter with the Dzunu<u>k</u>wa were his memories of it.

I met another Canadian Bigfoot researcher named Winona Kirk—she's married to John Kirk, the squatcher who said earlier that Bigfoot was as real as pizza is delicious. Winona said she'd talked to lots of First Nations elders and heard their stories. She'd also seen footprints near her home in Alberta. And one day, when she and her son were driving up the road, an animal crossed in front of them.

"What I thought in my head was orangutan," she told me. "A baby orangutan."

But this happened in late fall, in Alberta, Canada. The temperatures, especially at night, can get way below freezing! So it's not exactly the sort of place that orangutans—who usually make their homes in warm jungles, and on a completely different continent—would want to live. "And so I started thinking, yeah, that was probably a baby Bigfoot."

She didn't get a picture and she didn't know for sure, but it was definitely a strange moment, one that she thought about a lot. I enjoyed hearing Tom's and Winona's stories—they really made me wonder what they'd seen! Could it have been a Bigfoot? And there are hundreds more other stories out there.

Kathy Strain, one of the anthropologists I met, has been collecting hundreds of tribal Bigfoot stories that date back centuries. Some are very serious—clearly meant to teach the listener about the natural world and how to be safe in it. In those stories, being scared of Bigfoot wasn't necessarily a bad thing. In fact, running away from something could be a great way to stay alive. Other stories she collected are pretty funny. One of my favorites came from the Clackamas people in southern Oregon, who talked about whole clans of Bigfeet living together. In order to become an adult member of their Bigfoot tribe, an adolescent Bigfoot had to jump in front of a human on a trail and wave their hands in front of a human's face without being seen. Some of them were worse than others at doing this, which is why there have been Bigfoot sightings in that area.

Kathy wrote a whole book on Indigenous Bigfoot stories called *Giants, Cannibals and Monsters*. The stories came from all over the United States and Canada, from different tribes and at different times. Bigfoot was *everywhere*. Which made sense because long ago, there were more animals in general and more forests for them to live in. And because so many tribes have stories about a Bigfoot-like creature, it made me wonder if they had all caught sight of one of these mysterious animals. If you hear the same story from a lot of different people in different places, it makes you think it could perhaps be true.

For example, there are a lot of tales about giant floods from all over the world. There's one in the Bible and in the Quran, as well as others from ancient Mesopotamia—in parts of what are now Iraq and Turkey. And these flood stories show up in South America and Australia and were written about in ancient manuscripts in India. While portions of these stories don't seem realistic, it turns out that the flood part might actually have been true. Recently, geologists—scientists who study Earth and what it's made of—found proof that ten thousand years ago, when Earth was much cooler, enormous dams made of ice broke and caused huge floods. They also found evidence that when large meteors from space hit Earth's oceans, they caused giant waves and floods. Events like these might have been the reason there are so many stories about floods that have been passed down through generations. Maybe the same is true of Bigfoot! Of course, scientists have found the evidence they need to prove that parts of these flood stories could be true. We haven't yet found that evidence for Bigfoot.

I thought all these stories were fascinating, and I believed that many of the people who claimed to have seen Bigfoot did see something that they couldn't explain. But I could also understand why Grover wasn't sure what to make of these eyewitness encounters. Some of these stories are decades old, and people's memories shift, as the details of what happened get fuzzy. There's no real way of knowing how many of the sightings are credible—many are probably stumps, bears, other people, hoaxes, or just imaginations gone wild. Even the smartest people can mistake something they see in the woods for what they *want* to see. Scientists have a word for this: *pareidolia*. It's something that humans do a lot: find familiar patterns or shapes in something where they don't actually exist. The best example is when we think we see faces or animal shapes in rocks or clouds.

Even the stories that sounded completely believable still weren't the right kind of evidence—we know what we need (c'mon, DNA!). But like I've said before, we don't know for sure that Bigfoot *isn't* out there, so I think we should consider the possibility, the *maybe*. And these stories helped explain *why* people keep going out into the woods to solve a mystery that seems out of reach. In fact, all these stories about seeing a Bigfoot had me ready to go out and try to see one of my own!

GOIN' SQUATCHIN'

I'd been backpacking before and done a lot of hiking, but I decided I would need a little more help if I was going to go look for Bigfoot. Where should I go? What should I bring? Should I bring extra food for Bigfoot, just in case? How does Bigfoot feel about s'mores? Luckily, there are plenty of Bigfoot researchers who do this regularly and know just how to prepare. One of them is Cindy Caddell, who lives in Oregon and has been out searching for Bigfoot many, many times.

Cindy is an archaeologist—she studies the objects that our ancestors made, used, and left behind, so she can understand what their lives were like. She has worked with Native Americans to protect their ancient burial sites from wildfires. She's also a member of the

Olympic Project—the Bigfoot research group that has been keeping an eye on those mysterious nests. She sometimes leads expeditions into the woods to look for Bigfoot, and she agreed to take me! That meant I could call her up to ask her how to pack for this adventure!

"Make sure you have a tent, a sleeping pad, a pillow, and food and water," she said over the phone.

Well, that seems pretty normal, I thought to myself. *Nothing on that list seems very exciting or Bigfoot-specific.*

"And since we're going out in the dark, definitely bring a headlamp, since hiking at night can be kind of scary at first."

Uh-oh. I wrote down *headlamp* on my list. And extra batteries. Just in case.

She told me that sometimes people bring glow sticks and hang them from trees at night, because Bigfoot might be curious enough to come check them out. They also might pack special food for Bigfoot (I knew it!), hoping that if they left it out, they might attract one. I thought that Bigfoot might want some roots and vegetables, or some berries, or half of an elk (which was going to be hard to carry), but Cindy said she didn't do anything unusual. She sometimes brings some apples and will put them up high in a tree, where other animals won't find them as easily. She'd heard from others that if you're lucky, a Bigfoot might try to grab the apple and leave a nice impression of its big foot on the ground—although she'd never had that happen. I'd also heard from another squatcher that Bigfoot likes fruit pies—probably because they're sweet—but Cindy hadn't heard that before. And she made a really good point: Any food we brought would attract all kinds of animals, not just a Bigfoot. If your fruit pies and apples disappear in the night, it's just as likely to be raccoons or bears or squirrels. I decided to leave my pies at home (although I did nibble on one).

Cindy explained that, aside from the usual camping gear, I didn't really need to bring anything special. Especially because we'd have a couple other professional Bigfoot researchers joining us, and they'd have everything we might need. One of them was a guy named Gunnar Monson, and the other one, I was happy to learn, was Shane Corson! He was the guy who had led me to the nests, and since he's also a member of the Olympic Project, he and Cindy are good pals. We were all going to meet at a ranger station in the Mt. Hood National Forest, in Oregon. People have reported seeing Bigfoot in every state but Hawai'i (I guess because Bigfoot can't swim there!), but there have been hundreds of sightings in Oregon, especially in this area. The local highway is even called the Oregon Bigfoot Highway. If we were going to see Bigfoot anywhere, this seemed like it might be the right place.

I thanked Cindy for all her packing advice and hung up the phone. I'd be seeing her and Shane in just a few days, and I was super excited (and also a little nervous).

On a sunny but cool fall day, we all arrived at the ranger station where we'd planned to meet up. I was glad I'd brought a rain jacket and some extra-warm sweaters, because once the sun went down, it would feel pretty chilly. We hopped into our cars and followed one another up steep and winding roads that got narrower and less paved the farther into the mountains we went. The trees were thick, and the forest looked dark (and just a little spooky). We were going to a campsite that Shane had picked out—because he had had a possible Bigfoot encounter there once before.

Finally, we bounced over potholes into a dusty parking lot, where the trail to the campsite started. This was it! We were really doing this! I pulled my camping stuff out of the back of my very dirty car and hoped I hadn't brought too much. Then I looked over at Shane. He'd stuffed his backpack so full that I didn't know how he'd managed to fit his sleeping bag in there. I wasn't sure he'd even been able to squeeze in any food! Looked like I might be sharing with him instead of Bigfoot.

In his hand, Shane held three boxes that had a camouflage print on them. They were motion-sensing cameras, he told me. As we hiked to the campsite, he planned to go off the trail a little bit to strap them to a few different trees. If something walked by them, they'd snap a picture of it. He'd collect the cameras on our way back and see what animals were in the area. In the past, he'd gotten photos of mountain lions and bears, but no Bigfoot—at least not yet.

We hoisted our packs onto our backs and started off down the trail. Leaves crunched under our feet, but the forest was very quiet, aside from some birds chirping. Mossy pines towered overhead, and even though the sun shone brightly, it was still dark and cool under the canopy of leaves. Out of the corner of my eye, I caught a quick movement. *What was that?* I wondered. We'd been hiking for at least a

"Sometimes I wonder why we're doing this," Cindy said. "There's a lot of people that just drop out. They're tired of looking and it's not a guaranteed payoff."

The truth is that, for the most part, going squatchin' did not seem all that different from going camping. You get to tell ghost stories, go hiking in the woods, and hang out with friends. Cindy seemed to really enjoy that part. And she said she also still gets a thrill when she hears other people's Sasquatch stories—especially if they're people she never expected to talk about Bigfoot. Like her doctor, who once secretly told Cindy that she'd seen a Bigfoot cross a road right in front of her and her husband.

Cindy and a lot of other squatchers just want people to admit that Bigfoot *might* be possible, that it's not a completely crazy idea. And a lot of the people who are searching for Bigfoot are people who are interested in the natural world—the animals, the forests, the mountains. They want to know what's out there, and they have lots of questions, just like me. I think it's OK to ask those questions and wonder, "Maybe?" Bigfoot might not be out there, but it's worth exploring the world a little bit to find out.

We finished getting our tents set up and our sleeping bags unrolled. Clouds piled up over the mountains, and it started to get cold, so I pulled on a sweater and a wool hat. Then we carefully arranged a tiny teepee of sticks and twigs and dry leaves, before striking a match to light a fire. We added logs and pulled up some big rocks to sit on as we crowded around the dancing flames. Potatoes wrapped in aluminum foil roasted in the coals, and the last bits of sunlight faded from the sky.

That's when Shane decided it was time to tell us a story. Not just any story. A story about what happened the last time he camped here. He and his two friends had their tents in the same exact place, they built a fire in the same exact spot, and they had just finished eating their dinner and climbed into their sleeping bags. They were all sound asleep when, at about two in the morning, they started hearing this sound, like someone banging rocks together. They listened from inside their tents and could tell it was getting closer.

And closer.

AND CLOSER.

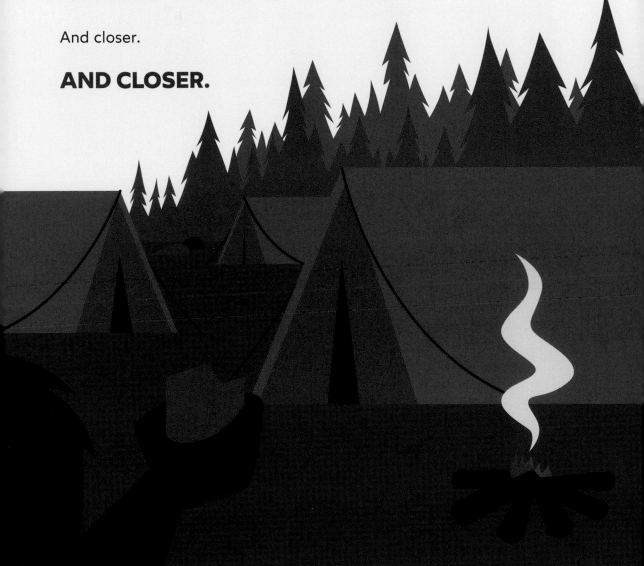

And then it stopped and went back up the trail in the direction it came from.

When they got up the next morning, they'd kinda forgotten about it. But that night, at the same time, it happened again. Something walking closer and closer down the trail, hitting rocks together, making a huge racket.

"It was crunching back and forth; it sounded as big as an elephant!" said Shane.

Then he pointed a finger to the hillside above our campsite, his face illuminated by the flickering firelight.

"Then, from up on that hill, we hear these loud knocks."

It sounded like someone cracking a baseball bat against a tree, as hard as they possibly could. Seconds later came a *thud*, like a rock hitting the ground just outside his tent.

"I just remember lying there. My heart was in my head," Shane said softly. "I was scared. I was so scared."

But he also wanted to know what was going on, so, bravely, he unzipped the nylon door of his tent and peered through the branches.

"That's when I saw this movement. Something was walking back and forth breaking branches." Shane pointed to a tree about twenty feet from our fire. "I could see a head and a shoulder and an arm. Then a hand. It was massive. And I don't know if it saw me or heard the sound of my zipper, but it took off into the darkness and was gone."

At dawn, Shane and his friends broke camp and hightailed it out of the woods, cutting their camping trip short. I don't know about you guys, but I don't think I would have even waited that long. I'm pretty sure I would have been packing everything up immediately. After I heard his story, I figured I wouldn't get much sleep, thinking about whatever monsters might be lurking in these woods. But, I found out, bedtime was still a long ways off, because no expedition would be complete without a nighttime excursion. I remembered now that during our phone call, Cindy had told me to bring a headlamp for just this reason—we were about to plunge into the woods, on our scary night hike.

A few minutes later, I fumbled with the button on my headlamp and switched it from ordinary white light to a setting that made it turn red. Red is harder to see from a distance and it's not so bright, so it's less likely to frighten any animals away, including, maybe, Bigfoot. I stretched the band of the headlamp over my woolly hat. I could see my breath in the red light, as it spiraled skyward in the cold. We started to move along the trail, up a steep, rocky hillside. I tried to be quiet, but it was hard to see, and I tripped over some loose rocks. Whoops! Probably better to go a little slower than to take a tumble. I paid closer attention to where I put my feet and followed Shane, Cindy, and Gunnar. After climbing and climbing, we reached a level point in the path and

"No, turn toward the camp and then walk straight."

The thing still didn't move, but Shane could see that Gunnar glowed much brighter through the FLIR goggles than whatever the white thing was, which meant he was the warmest thing in the area.

"I'm thinking it's a tree," Cindy sighed.

Wait. A tree? Yep! Cindy said that if the tree had something like bacteria or ants on it, those give off heat and would be picked up by the FLIR.

"Sorry about that," she said. "But it *did* look like a head and shoulders."

I felt way more disappointed than I had expected. In the moment, I'd almost forgotten to be scared. Shane made another few calls out into the darkness. But we were all a little worn out by the excitement. So we made our way back to the camp, our tents, and the warmth of our sleeping bags. No luck, but I still held out hope for the results of those DNA tests.

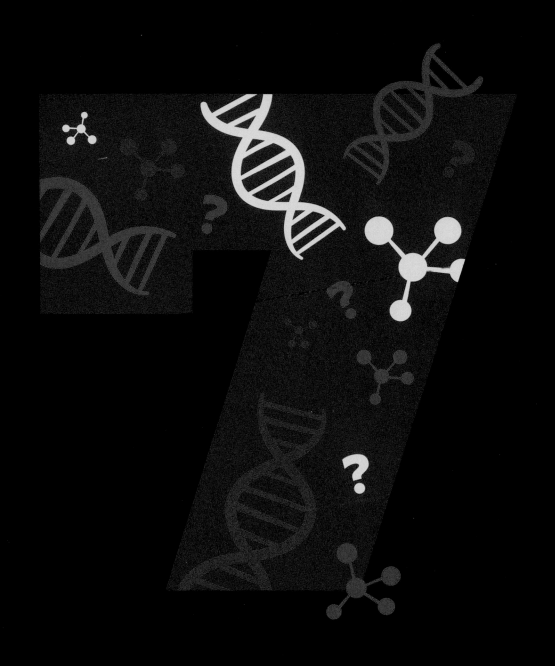

SCIENCE OR FICTION?

Ding ding!

When I returned to civilization after being in the woods, my cell phone finally picked up a signal. A whole bunch of texts and emails came flooding in, as well as the little bells that told me I had a voice mail.

It's probably my parents, I thought. I'd told them where I was going and when I'd be back, but I knew they'd been a little worried that I'd be eaten by something while I was out in the woods. I started to listen to the message and almost dropped the phone in excitement.

"Hey, Laura—it's Dr. Disotell. I have the results from the DNA analysis, if you want to give me a call."

I couldn't tell from his voice on the message he'd left if it was good news or bad news and started to get really nervous about what he'd found—or hadn't. It was already too late to call him back, and anyway, he wouldn't have been at his lab.

But the next morning, bright and early, I was up and dialing his phone number as soon as I could.

What if those nests are proof of Bigfoot? I thought. *But what if they aren't?*

My stomach jumped with butterflies as the phone rang and rang and rang, and just as I thought it would go to voice mail, Dr. Disotell picked up. I could barely get out a polite "Hello, how are you?" before I jumped right into the question I was dying to know the answer to.

"What did you find?"

"Well," he responded, "we looked at the DNA sequences taken from the nest samples that the Olympic Project collected and sent to me. But the problem is that they're so old that they've degraded a little— they aren't in great condition. Nevertheless, we still got millions of individual DNA sequences."

Millions! I thought. *There's got to be something good in there!*

"And?" I asked, trying to stay calm.

"And . . . the only primate we found was human. A specific real human, not just something close to human. No unknown primates."

My heart sank. This was NOT what I wanted to hear.

"Was there anything weird at all? Anything that didn't make sense, or you couldn't figure out what it was?" I asked, hoping against hope.

"Let's see . . . going down my list here . . . I got some bats, some shrews, and lots of bear DNA. Also some skunks, weasels, and rabbits."

I wasn't surprised that the test came up with all these different animals—that's what happens with eDNA. But, oh man. I was super disappointed. I didn't realize how much I wanted there to be something that might hint at the possibility of Bigfoot! And if there wasn't anything unusual in the DNA, then what made those huge nests, deep in the woods? They were so, so bizarre! And the only results we got from all that weirdness were for bats and shrews and bears? No Bigfoot?

Just as I was about to tell Dr. Disotell thank you and goodbye, he said he did find one thing that seemed out of place: horse DNA.

That is *strange*, I thought.

The area where the nests were discovered isn't exactly the kind of place where you'd find horses—it's too steep and overgrown with shrubs. Dr. Disotell thought this might mean someone stepped in horse manure—or crossed a field where horses spent time—and tracked that mud or poop into the nest site. Then the DNA from the horses would have been mixed in with all the other DNA. Dr. Disotell calls this contamination, meaning that horse DNA made the nest samples dirty.

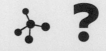

Now wait, you might be thinking. Those samples were filled with sticks and leaves and mud—not exactly clean to begin with! What Dr. Disotell meant by "dirty" is that there's DNA you would expect to find in an area, as well as outside DNA that shouldn't be there at all. When the people who are collecting samples from somewhere are doing it correctly, they keep that outside DNA away by wearing gloves, hairnets, and booties over their shoes. Sometimes, they'll even wear a white zip-up suit over all their clothes—you might have seen this in television shows or the movies. They're specifically trying to avoid contaminating the area they're working in with things that shouldn't be there. So the presence of horse DNA—which shouldn't have been there—made Dr. Disotell think the nest site was contaminated.

But that, he said, was the only odd thing about the samples.

It was kind of a letdown. And if I was feeling down, I could only imagine how the members of the Olympic Project and Dr. Jeff Meldrum felt. He'd spent so much time collecting the nest samples, packaging them up to send to Dr. Disotell's lab, and then waiting for the results! He must have been super bummed out!

But when I called him to ask, he didn't seem too upset.

"You know, I've learned to steel myself against negative results over the years," he said. "There are just so many times that a scientist thinks they have something that turns out to be nothing. But also, I'm thinking about what Dr. Disotell said about the DNA—that it was old and degraded."

He pointed out that the nests sat out in the forest while rain poured down on them, animals and people came tromping through, and the summer sun cooked them—all before anyone took any samples. So if someone could find better nests? Newer ones? Then maybe we might get better information!

Dr. Meldrum also explained that we don't know what Bigfoot is and where it fits on the family tree. Is it more closely related to great apes? Or is it more closely related to humans? What if Bigfoot DNA and human DNA are very, very similar, and the samples he collected weren't good enough to tell the difference? Dr. Meldrum said he hoped to go back out to that same area again soon and explore a little deeper into the woods, to see what might turn up. I crossed my fingers that he'd find something interesting and do some more tests.

But in the back of my mind, I also realized that if a Bigfoot, or a herd of them, had built those nests and stayed there for a while, then that whole area would be absolutely swimming with Bigfoot DNA and other proof of its existence.

I'd heard some other ideas about why it's so hard to get the evidence we need. They came from a group of Bigfoot seekers who don't think physical evidence—hair, DNA, footprints—are important. Instead, they had more magical, less scientific explanations. For them, Bigfoot isn't necessarily flesh and blood, like all the other creatures on this planet. Their theories can get pretty wild. I might have mentioned a few of them before—like the idea that Bigfoot can make itself invisible. Some people I spoke with said Bigfoot has some sort of superpower that helps it disappear, kind of like Harry Potter and his cloak of invisibility. Another guy I spoke to suggested that Bigfoot could teleport—move from one place to another instantly. One minute Bigfoot's standing there, and then—*poof*—it's gone. I'd also heard that Bigfoot is a visitor from another planet and that aliens dropped Bigfoot off with their spaceship.

These are definitely fun and imaginative ideas, but they're more science fiction than science. No other creature on this planet can teleport or become invisible. There are some animals like lizards and fish that are really good at blending into their surroundings through camouflage—and maybe Bigfoot could do something like that!—but it doesn't seem very likely that Bigfoot turns invisible.

In my opinion, if we want to be scientific about this search, we have to apply the same rules of science to Bigfoot that we do to everything else. So for the serious Sasquatch researchers I talked to, the idea that Bigfoot is a shape-shifting, time-traveling, invisible alien from another planet drove them a little crazy. They referred to these more fantastical ideas as "the Woo."

The Woo shows up in corny television shows where Bigfoot has all those magical powers I just mentioned. Or it's in those silly tabloid newspapers, the ones you see in the grocery store checkout line, with stories about how Bigfoot lives in someone's basement or that Bigfoot is running for president. They're funny to read, but they also make Bigfoot seem like a joke. Every time a silly or magical story about Bigfoot comes out, science-minded researchers worry they're not going to be taken seriously. Even Grover knew this was a problem.

"It makes the whole field look silly," he once said in an interview. "And the serious researchers tend to back off. It makes you look bad in the eyes of your peers and superiors." As we've learned, Grover approached the Bigfoot question with the idea that this was a flesh-and-blood creature that didn't act differently from any other creature on the planet. He looked at the evidence and knew we didn't have enough to prove Bigfoot's existence, but he wanted to keep looking by using science and logic. And because he was a real scientist and professor, a lot of people respected him and wanted his opinion on their Bigfoot theories, including the "Woo" ones.

Grover would go to meetings or be interviewed as a guest on radio shows, and he would want to talk about the science he was using to look for Bigfoot. But he would get questions from people who asked if Bigfoot was made of titanium, or who would tell Grover that they'd seen a Bigfoot disappear in front of their eyes. Grover listened politely because he didn't want to be rude to people, but he was very frustrated by this.

He didn't really care if these ideas made him look bad, but he knew that other scientists might not want to look for Bigfoot because they'd get made fun of or maybe have trouble at work. A lot of Grover's work colleagues didn't believe in Sasquatch or even think it was something Grover should be spending time on. They were embarrassed by his work on Bigfoot and made jokes about him. Grover loved his work, and he loved teaching. Even though he had a pretty thick skin and didn't worry about what people thought, he knew that other scientists might not like being made fun of and would stay away from Bigfoot entirely.

That bothered Grover because he wanted more scientists doing real science to think about the Bigfoot question and help him solve it. He wanted scientific answers to a scientific question. And while a lot of people think this Sasquatch question is silly, there are a lot of other crazy questions that scientists *do* take seriously. Questions like: Is there intelligent life on other planets? Can plants and trees talk? Is the universe real or just a computer program? Scientists are asking all sorts of weird questions, and a lot of them seem pretty "Woo." So I think it's OK that we ask questions about Bigfoot, too. Questions are a big part of science, of understanding the world around us. This is where the scientific method becomes important: asking a clear question, forming a solid hypothesis, designing an experiment to test the hypothesis, gathering the data, and analyzing your results to see if your hypothesis was correct.

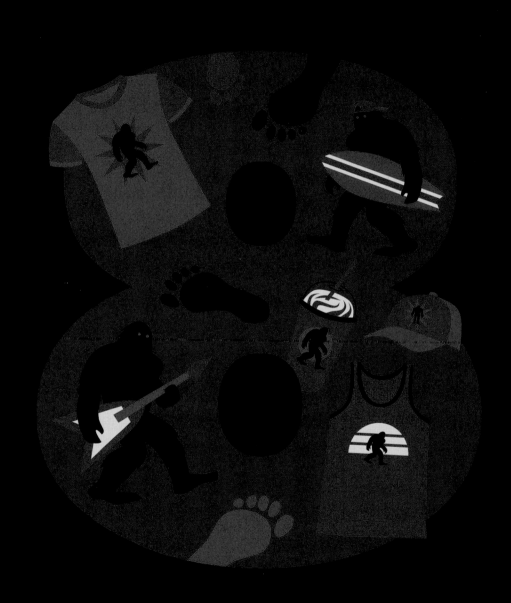

WHY WE WANT TO BELIEVE

Just about everyone I know has heard of Bigfoot, and it's no surprise. Once you start looking around, Bigfoot is everywhere. Bigfoot Tires, Bigfoot Bikes, Sasquatch Bikes, Bigfoot Coffee, Sasquatch Chocolate, Bigfoot Breath Mints, Sasquatch Snowboards, Sasquatch Books, Bigfoot Ale, the Sasquatch! Music Festival. Whew! And that's just some of them—there are also thousands of books, movies, and television shows about Bigfoot, as well as Bigfoot video games and comic books and even Bigfoot museums!

Not far from where I live in Denver, Colorado, there's a little town up in the mountains called Bailey. It's home to the Sasquatch Outpost pleased by how many people he had inspired to look for Bigfoot, and to do so using science. He always loved teaching, and his work lives on in the next generation of Bigfoot researchers, people like Shane Corson. I talked with Shane after the results of the DNA tests came back. Even though he felt disappointed, he's not ready to give up.

"The bottom line is something is making these nests," he said. "And they need to be properly documented and researched thoroughly."

So Shane still goes out in the woods, to see what he can find.

"It's something I really enjoy doing. I get to be involved in nature, and there's that chance of discovery, and maybe being involved in that discovery."

Lots of the other squatchers I talked to hope for the same thing.

What about me? I heard some hair-raising stories about seeing or hearing or running into a Bigfoot. I saw some giant plaster footprint casts and bits of unknown hair. I watched the Patterson-Gimlin film over and over. I looked at lots of blobsquatches. But I didn't see that one piece of evidence that we absolutely need to prove Bigfoot exists—a real live Bigfoot (or at least its DNA). Even so, I still can't say for sure it's not there. So I'm going to keep asking questions and considering the different possibilities (using the scientific method, of course). And remember, no one has proven that Bigfoot *doesn't* exist. So I'll probably keep my eyes open, just in case—and you should, too.

BIGFOOT EXPEDITION PACKING LIST

So you want to go out on your own Bigfoot expedition! Well, first, you'll have to find a responsible adult to go with you—someone who can help make sure you have everything you need, set up tents, and build fires (and fire safety is very important—make sure you find out what the rules are for campfires where you're going). And you'll probably need that same responsible adult to drive! You don't have to go deep into the woods to try to see Bigfoot—a lot of people claim to have seen Sasquatch much closer to civilization.

And no matter if you're going way into the wilderness or just to your nearby park, if you're going just for the day, for one night, or for a whole week, you need to tell someone where you are going, where you are planning to camp, and when you are coming back.

Now, a few things you should think about before you start packing:

1. **Where are you going? Will there be lots of rocky, steep trails or is it a nice, grassy meadow?**
2. **What will the weather be like? Hot and sunny? Cold and windy? Wet?**

These questions will help you pack the right kinds of gear for your adventure. And now, a basic packing list!

Stuff you need, no matter what:

- The right kind of clothes for the weather—will it be cold? Rainy? Hot?
- Good walking shoes or boots
- Map and compass
- First aid kit
- Food—don't forget fruit pies, just in case!
- Plenty of water
- Headlamp or flashlight
- A phone, radio, or whistle to call for help in an emergency

Stuff you need if you're camping:

- Tent
- Sleeping bag
- Sleeping pad
- Matches or a lighter, fire starter (like cotton balls coated in Vaseline), and a responsible adult!

Stuff you might want to have:

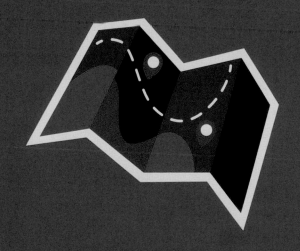

- Sunglasses and sunscreen
- Bug spray
- Camp chair
- Glow sticks
- Spooky stories

Stuff specifically for squatchin':

- Notebook and something to write with—to take notes about what you see/hear/smell, the weather, where you go, what kinds of plants are around
- Tape measure—for measuring footprints if you find any
- Camera—to get blurry blobsquatch photos
- Binoculars
- Plastic, resealable bags—for collecting samples
- Latex gloves and tweezers—so you don't contaminate your samples

Stuff to do while you're out there:

- Look for footprints, bits of hair, broken twigs, and nests. Where are they? Are there lots of them? What do they look like? What might have made them or left them behind?
- Take photos of what you find and of the environment you're in, as well as any Sasquatches you see (and get some of you and your responsible adult, too!).
- Listen for grunts or odd sounds. What might have made those noises? What else do you hear?
- Tell scary stories (but not too scary).
- Make sure you clean up your campsite, put out your fire completely (every last ember), and leave everything better than you found it. Be like Bigfoot—leave no trace!

BIGFOOT'S COUSINS AROUND THE WORLD

Like Bigfoot, none of these creatures have been scientifically proven to exist.

MAPINGUARY (BRAZIL)—This giant, hairy (and very smelly!) creature supposedly lives in the Amazon rain forest. Some people claim it only has one eye, like the Cyclops, although others say it has two. It's also rumored to have a big mouth in its belly! Some scientists think that legends about this creature are based on sightings of giant ground sloths, which are now thought to be extinct.

ORANG PENDEK (INDONESIA)—The shortest of Bigfoot's cousins, this three-to-five-feet-tall creature is thought to live in the forests of Sumatra, an island in the Indian Ocean. Eyewitnesses describe a very hairy, apelike thing that walks on two legs and has a face like a human. Several scientific expeditions have gone looking for the Orang Pendek, thinking that it might be an unknown species of orangutan, or perhaps a relative of *Homo floresiensis*.

YĚRÉN (CHINA)—Like its North American cousin, this long-haired creature reportedly walks on two legs, although it's thought to be a little shorter (only six feet tall). Stories about the Yěrén go back hundreds of years. And, like Bigfoot, some people think the Yěrén might be descended from *Gigantopithecus*!

YETI (NEPAL and BHUTAN)—Stories about the Yeti, aka the Abominable Snowman, go back hundreds of years, but the creature became world-famous when Europeans exploring the Himalayan mountains began sharing tales of strange sightings and tracks. In recent years, DNA studies conducted on supposed Yeti hair suggest that the creature may actually be a species of brown bear.

YOWIE (AUSTRALIA)—Bigfoot's cousin Down Under! This six (but maybe twelve?)-foot-tall creature supposedly has reddish-brown hair and only four toes, and it lives in the Outback. Stories of the Yowie—which have their roots in Aboriginal legends—may be based on a now-extinct Australian ape known as the Yahoo.

Mapinguary

Orang Pendek

Yeti

Yěrén

Yowie

GLOSSARY

anatomy—The study of organisms and how they're put together and organized.

Anglicize—To adapt a word in a different language into English.

anthropology—The study of humans, both past and present. Physical anthropology looks at how humans evolved from other animals and how they have adapted. Cultural anthropology looks at how people live, like the rules they follow and what values and ideas they consider important. An anthropologist is someone who studies anthropology.

archaeology—The study of humans by looking at the things they've left behind, like art and tools and pottery. Someone who studies archaeology is an archaeologist.

bases—The chemicals that make up DNA: adenine, thymine, guanine, and cytosine. Two bases joined together make up a base pair, and these are the "rungs "in the ladder that make up the structure of DNA.

belief—The idea that something is true but can't be proven by science.

bipedal—Walks on two legs.

circumstantial evidence—Evidence that seems to lead to a conclusion but may not actually prove it as a fact.

classification—The act of arranging similar things into groups.

Coast Salish—A group of Native Americans and First Nations Indigenous peoples who live along the Pacific coastline from British Columbia, in Canada, down to Oregon.

conservation—The protection of things found in nature.

contaminate—Making something impure or unusable by adding something that pollutes or dirties it.

cryptozoology—The search for unknown and unproven creatures.

Denisovan—A now-extinct relative of modern humans, whose fossil remains were found by scientists in 2010 in a cave in Siberia.

direct evidence—Evidence that proves a conclusion.

double helix—The description of the structure of a molecule of DNA, which looks like a twisted ladder.

environmental DNA—DNA that is collected from the environment where a creature may have spent time, rather than collecting the DNA from the actual creature.

evolution—A scientific theory that explains how living things change over time and become new species.

fossil—Hardened remains or traces of organisms that lived long ago.

fossilized—To be changed into a fossil; to be preserved in a hardened form.

gene—The basic unit of heredity that contains information about how an organism looks and functions.

geology—The study of the Earth and its physical features. Scientists who study geology are called geologists.

hominid—All the members of the biological family Hominidae. This includes humans, chimpanzees, gorillas, and orangutans, as well as all our extinct relatives.

hypothesis—An educated guess that can be tested by science.

Indigenous—The groups of people who were the first known to live in any region. In North America, this refers to Native American and First Nations peoples.

iteration—A step where an action is repeated, in this case as part of testing a hypothesis.

lepidopterist—A scientist who studies butterflies and moths.

mnemonic—A trick for helping people remember information more easily.

mutation—A mistake or change in an organism's DNA.

objective—Looking at the facts without letting personal feelings get in the way.

paleontology—The study of ancient life. A scientist who studies paleontology is a paleontologist.

pareidolia—Finding familiar patterns or shapes in something where they don't actually exist.

primate—An animal in the category of mammals that includes humans, monkeys, apes, and lemurs.

scientific method—A research process that uses a set of steps to make observations, collect information, and draw conclusions about a scientific question.

sequence—The order of letters A, T, G, and C that make up a molecule of DNA.

species—A group of organisms that are closely related to one another; they share similar characteristics, and they can reproduce with each other.

specimen—A sample of something. A type specimen is the original specimen from which a new species is identified.

trait—A physical characteristic, like hair or eye color.

NOTES

INTRODUCTION: NESTS

2 Description of Bigfoot: Grover S. Krantz, *Big Footprints: A Scientific Inquiry into the Reality of Sasquatch* (Boulder, CO: Johnson, 1992), 2.

5 Gorilla nests: See "#GorillaStory: Nest Building," Smithsonian's National Zoo & Conservation Biology Institute, January 26, 2018, nationalzoo .si.edu/animals/news/gorillastory-nest-building.

6 Sasquatch and Coast Salish: See Dorothy Kennedy and Randy Bouchard, "Coast Salish," *Canadian Encyclopedia*, February 7, 2006 (updated July 25, 2019), www.thecanadianencyclopedia.ca/en/article/coastal-salish.

6 Bigfoot's name: See en.wikipedia.org/wiki/Bigfoot#Regional_names.

CHAPTER 1: GROVER

12 What is anthropology: See "What Is Anthropology?," *American Anthropological Association,* www.americananthro.org/Advance YourCareer/Content.aspx?ItemNumber=2150.

13 Grover Krantz Smithsonian exhibit caption: See Haleema Shah, "The Scientist Grover Krantz Risked It All . . . Chasing Bigfoot," *Smithsonian*, October 31, 2018, www.smithsonianmag.com/smithsonian-institution /scientist-grover-krantz-risked-it-all-chasing-bigfoot-180970676.

13 Grover Krantz in the *Washington Post*: See Peter Carlson, "Using His Cranium Grover Krantz's Last Wish Was to Remain with His Friends. And He Has," *Washington Post*, July 5, 2006, www.washingtonpost

.com/archive/lifestyle/2006/07/05/using-his-cranium-span
-classbankheadgrover-krantzs-last-wish-was-to-remain-with-his-friends
-and-he-hasspan/2d856129-1c26-49ae-a0df-8726c435f70a.

16 "Bigfoot is a large": Grover Krantz, "Big Footprints," interview
by John Yager, KXLY-TV, Spokane, WA, 1992, www.youtube.com
/watch?v=Rm_15lJ6z70, 7:26.

18 Bigfoot likes doughnuts: See David Moye, "Sasquatch's Favorite Foods
Revealed by 'Finding Bigfoot' Star Bobo Fay," *HuffPost*, November 7,
2013, www.huffpost.com/entry/bigfoot-bobo-fay_n_4234898.

20 "If someone faked that print": Krantz, Yager interview, 9:10.

CHAPTER 2: **WE ARE FAMILY**

28 Evolution explainer: See Britannica Kids, kids.britannica.com/students
/article/evolution/274236.

33 Human classification: See Maggie A. Norris and Donna Rae Siegfried,
"Taxonomy of Homo Sapiens," *Dummies*, www.dummies.com
/education/science/anatomy/taxonomy-homo-sapiens.

34 Lucy: See Lisa Hendry, "*Australopithecus Afarensis*, Lucy's
Species," *Natural History Museum*, www.nhm.ac.uk/discover
/australopithecus-afarensis-lucy-species.html.

35 Neanderthals: See "*Homo Neanderthalensis*," *Smithsonian National
Museum of Natural History*, humanorigins.si.edu/evidence/human
-fossils/species/homo-neanderthalensis.

37 *Homo floresiensis*: See Karen L. Baab, "*Homo Floresiensis*:
Making Sense of the Small-Bodied Hominin Fossils from Flores,"
Nature Education: Knowledge Project, 2012, www.nature.com
/scitable/knowledge/library/homo-floresiensis-making-sense-of-the
-small-91387735.

38 Denisovans: See "Denisovans: The Ancient Humans Who Share Our
Ancestry," *New Scientist*, www.newscientist.com/definition/denisovans.

38 *Homo naledi*: See Colin Barras, "*Homo Naledi* Is Only 250,000 Years Old—Here's Why That Matters," *New Scientist*, April 25, 2017, www.newscientist.com/article/2128834-homo-naledi-is-only-250000 -years-old-heres-why-that-matters.

39 *Gigantopithecus*: See en.wikipedia.org/wiki/Gigantopithecus#Size.

40 "Here I was now": Paul Du Chaillu, *Explorations and Adventures in Equatorial Africa* (London: Murray, 1861), 58.

40 "Nearly six feet high": Ibid., 70.

40 Gorilla: See en.wikipedia.org/wiki/Gorilla.

42 Gorilla ballet: See Richard Conniff, "The Missionary and the Gorilla," *Yale Alumni Magazine*, September/October 2008, yalealumnimagazine .com/articles/2182-the-missionary-and-the-gorilla?page=5.

CHAPTER 3: THE EVIDENCE

48 "The way I like": Krantz, Yager interview, 7:12.

48 Ray Wallace: See Scott Martelle, "Ray Wallace, 84; Took Bigfoot Secret to Grave—Now His Kids Spill It," *Los Angeles Times*, December 6, 2002, www.latimes.com/archives/la-xpm-2002-dec-06-me -wallace6-story.html; Timothy Egan, "The Search for Bigfoot Outlives the Man Who Created Him," *New York Times*, January 3, 2003, www.nytimes.com/2003/01/03/us/search-for-bigfoot-outlives-the -man-who-created-him.html.

52 Patterson-Gimlin: See en.wikipedia.org/wiki/Patterson% E2%80%93Gimlin_film.

55 "The walk is peculiar": Grover Krantz, "Big Footprints: A Scientific Inquiry," interview by Bob Hieronimus, Hieronimus & Co., *21st Century Radio*, January 17, 1993, 23:18.

55 Grover walking: See www.youtube.com/watch?v=6ocBNs8sl_g, 2:50.

55 "Almost any scientist": Krantz, Yager interview, 3:50.

56 "Someone might want 'final proof'": Series 5: Sasquatch, Grover Sanders Krantz Collection, National Anthropological Archives, Smithsonian Institution, sova.si.edu/record/NAA.2003-21?s=0&n=10&t=C&q=&i=0.

CHAPTER 4: **BLUEPRINTS, NOT FOOTPRINTS**

63 DNA explainer: See Britannica Kids, kids.britannica.com/kids/article /DNA/390730.

63 Water flea: See Mark Brown, "Tiny Water Flea Has More Genes Than You Do," *Wired*, February 4, 2011, www.wired.com/2011/02/water -flea-genome.

65 "There are some new techniques": Krantz, Hieronimus interview, 30:15.

66 Right kind of environment: See Edna Sadayo Miazato Iwamura,, José Arnaldo Soares-Vieira, and Daniel Romero Muñoz, "Human Identification and Analysis of DNA in Bones," *SciELO,* 2004, www.scielo.br/j/rhc/a /g4jn3MBK8hxYSpYGP5vJGYw/?lang=en.

66 Bones and teeth: "Sources of DNA," *Biology Project: Human Biology.* See www.biology.arizona.edu/human_bio/problem_sets/dna _forensics_2/06t.html.

70 Yeti DNA: See Sid Perkins, "So Much for the Abominable Snowman: Study Finds That 'Yeti' DNA Belongs to Bears," *Science,* November 28, 2017, www.sciencemag.org/news/2017/11/so-much-abominable -snowman-study-finds-yeti-dna-belongs-bears.

70 Loch Ness Monster DNA: See "Loch Ness Monsters May Be a Giant Eel, Say Scientists," *BBC,* September 5, 2019, www.bbc.com/news /uk-scotland-highlands-islands-49495145.

CHAPTER 5: **EYEWITNESS**

85 Clackamas: Kathy Moskowitz Strain, *Giants, Cannibals and Monsters: Bigfoot in Native Culture* (Blaine, WA: Hancock House, 2008), 152.

86 Flood myths: See Lennlee Keep, "A Flood of Myths and Stories," *Independent Lens for PBS*, February 14, 2020, www.pbs.org /independentlens/blog/a-flood-of-myths-and-stories.

86 Ice dam: See David R. Montgomery, "Biblical-Type Floods Are Real, and They're Absolutely Enormous," *Discover*, August 28, 2012, www. discovermagazine.com/planet-earth/biblical-type-floods-are -real-and-theyre-absolutely-enormous.

86 Pareidolia: See Larry Sessions, "Seeing Things That Aren't There? It's Called Pareidolia," *EarthSky*, November 25, 2020, earthsky.org /human-world/seeing-things-that-arent-there.

CHAPTER 7: **SCIENCE OR FICTION?**

112 "It makes the whole field": Krantz, Hieronimus interview, 54:03.

114 Scientific method: See "Steps of the Scientific Method," *Science Buddies*, www.sciencebuddies.org/science-fair-projects/science-fair /steps-of-the-scientific-method.

CHAPTER 8: **WHY WE WANT TO BELIEVE**

125 "I want it settled now": Grover Krantz, "Sasquatch: The Anthropology of the Unknown," *The Anthropology of the Unknown: Sasquatch and Similar Phenomena*, May 10, 1978, Radio Canada International, Montreal, disc 5, 14:04.

125 Looking until 2001: Series 5: Sasquatch.

BIGFOOT'S COUSINS AROUND THE WORLD

131 Mapinguary: See Larry Rohter, "A Huge Amazon Monster Is Only a Myth. Or Is It?" *New York Times*, July 8, 2007, www.nytimes .com/2007/07/08/world/americas/08amazon.html.

131 Orang Pendek: See "Orang Pendek: A Cryptozoological Animal in the Sumatran Wilderness," *Wild Sumatra,* www.wildsumatra.com /orang-pendek.

131 Yĕrén: See Travis Smola, "5 Bigfoot Legends from Around the World That May Actually Be Real," *Wide Open Spaces,* December 23, 2020, www.wideopenspaces.com/5-bigfoot-legends-from-around-the-world -that-may-actually-be-real.

132 Yeti: See Perkins, "So Much for the Abominable Snowman."

132 Yowie: See en.wikipedia.org/wiki/Yowie.

BIBLIOGRAPHY

ARCHIVES

Grover Sanders Krantz Collection. National Anthropological Archives, Smithsonian Institution. See sova.si.edu/record/NAA.2003 -21?s=0&n=10&t=C&q=&i=0.

BOOKS

Du Chaillu, Paul. *Explorations and Adventures in Equatorial Africa.* London: Murray, 1861.

Gordon, David George. *The Sasquatch Seeker's Field Manual.* Seattle: Mountaineers Books, 2015.

Krantz, Grover S. *Big Footprints: A Scientific Inquiry into the Reality of Sasquatch.* Boulder, CO: Johnson, 1992.

Strain, Kathy Moskowitz. *Giants, Cannibals and Monsters: Bigfoot in Native Culture.* Blaine, WA: Hancock House, 2008.

LECTURES

Krantz, Grover. "Sasquatch: The Anthropology of the Unknown." *The Anthropology of the Unknown: Sasquatch and Similar Phenomena.* May 10, 1978, Radio Canada International, Montreal.

PERIODICALS

Barras, Colin. "*Homo Naledi* Is Only 250,000 Years Old—Here's Why That Matters." *New Scientist,* April 25, 2017. See www.newscientist.com

/article/2128834-homo-naledi-is-only-250000-years-old-heres-why
-that-matters.

Brown, Mark. "Tiny Water Flea Has More Genes Than You Do." *Wired*,
February 4, 2011. See www.wired.com/2011/02/water-flea-genome.

Carlson, Peter. "Using His Cranium Grover Krantz's Last Wish Was to
Remain with His Friends. And He Has." *Washington Post*. July 5, 2006.
See www.washingtonpost.com/archive/lifestyle/2006/07/05/using
-his-cranium-span-classbankheadgrover-krantzs-last-wish-was-to
-remain-with-his-friends-and-he-hasspan/2d856129-1c26-49ae-a0df
-8726c435f70a.

Conniff, Richard. "The Missionary and the Gorilla." *Yale Alumni
Magazine*, September/October 2008. See yalealumnimagazine.com
/articles/2182-the-missionary-and-the-gorilla?page=5.

Egan, Timothy. "The Search for Bigfoot Outlives the Man Who Created Him."
New York Times, January 3, 2003. See www.nytimes.com/2003/01/03
/us/search-for-bigfoot-outlives-the-man-who-created-him.html.

Martelle, Scott. "Ray Wallace, 84; Took Bigfoot Secret to Grave—Now
His Kids Spill It." *Los Angeles Times*, December 6, 2002. See www.
latimes.com/archives/la-xpm-2002-dec-06-me-wallace6-story.html.

Montgomery, David R. "Biblical-Type Floods Are Real, and They're
Absolutely Enormous." *Discover*, August 28, 2012. See www.
discovermagazine.com/planet-earth/biblical-type-floods-are
-real-and-theyre-absolutely-enormous.

Perkins, Sid. "So Much for the Abominable Snowman: Study Finds
That 'Yeti' DNA Belongs to Bears." *Science*, November 28, 2017. See
www.sciencemag.org/news/2017/11/so-much-abominable-snowman
-study-finds-yeti-dna-belongs-bears.

Rohter, Larry. "A Huge Amazon Monster Is Only a Myth. Or Is It?" *New
York Times*, July 8, 2007. See www.nytimes.com/2007/07/08/world
/americas/08amazon.html.

Shah, Haleema. "The Scientist Grover Krantz Risked It All . . . Chasing Bigfoot." *Smithsonian*, October 31, 2018. See www.smithsonianmag.com /smithsonian-institution/scientist-grover-krantz-risked-it-all-chasing -bigfoot-180970676.

RADIO AND TELEVISION INTERVIEWS

Krantz, Grover. "Big Footprints." Interview. *Discovery Channel*. I was unable to find the original footage, but excerpts of it can be found here: www.youtube.com/watch?v=6ocBNs8sl_g.

Krantz, Grover. "Big Footprints." Interview by reporter John Yager. KXLY-TV, Spokane, WA, 1992. The video was never broadcast but circulated within the Bigfoot community. I found segments of it online here: www.youtube.com/watch?v=Rm_15lJ6z70.

Krantz, Grover. "Big Footprints: A Scientific Inquiry." Interview by Bob Hieronimus, Hieronimus & Co. *21st Century Radio*, January 17, 1993.

WEBSITES

"GorillaStory: Nest Building." "#GorillaStory: Nest Building." Smithsonian's National Zoo & Conservation Biology Institute, January 26, 2018. See nationalzoo.si.edu/animals/news/gorillastory-nest-building.

Baab, Karen L. "*Homo Floresiensis:* Making Sense of the Small-Bodied Hominin Fossils from Flores." *Nature Education: Knowledge Project,* 2012. See www.nature.com/scitable/knowledge/library/homo-floresiensis -making-sense-of-the-small-91387735.

Bigfoot. See en.wikipedia.org/w/index.php?title=Bigfoot&o ldid=1032486168.

"Denisovans: The Ancient Humans Who Share Our Ancestry." *New Scientist.* See www.newscientist.com/definition/denisovans.

"DNA." *Britannica Kids.* See kids.britannica.com/kids/article/DNA/390730.

"Evolution." *Britannica Kids*. See kids.britannica.com/students/article /evolution/274236.

Gigantopithecus. See en.wikipedia.org/wiki/Gigantopithecus#Size.

Gorilla. See en.wikipedia.org/wiki/Gorilla.

Hendry, Lisa. "*Australopithecus Afarensis*, Lucy's Species." *Natural History Museum*. See www.nhm.ac.uk/discover/australopithecus-afarensis -lucy-species.html.

"*Homo Neanderthalensis*." *Smithsonian National Museum of Natural History*. See humanorigins.si.edu/evidence/human-fossils/species /homo-neanderthalensis.

Iwamura, Edna Sadayo Miazato, José Arnaldo Soares-Vieira, and Daniel Romero Muñoz. "Human Identification and Analysis of DNA in Bones." *SciELO*, 2004. See www.scielo.br/j/rhc/a/g4jn3MBK8hxYSpYGP5 vJGw/?lang=en.

Keep, Lennlee. "A Flood of Myths and Stories." *Independent Lens for PBS*, February 14, 2020. See www.pbs.org/independentlens/blog/a-flood-of -myths-and-stories.

Kennedy, Dorothy, and Randy Bouchard. "Coast Salish." *Canadian Encyclopedia*, February 7, 2006 (updated July 25, 2019). See www.the canadianencyclopedia.ca/en/article/coastal-salish.

"Loch Ness Monsters May Be a Giant Eel, Say Scientists." *BBC*, September 5, 2019. See www.bbc.com/news/uk-scotland-highlands -islands-49495145.

Moye, David. "Sasquatch's Favorite Foods Revealed by 'Finding Bigfoot' Star Bobo Fay." *HuffPost*, November 7, 2013. See www.huffpost.com /entry/bigfoot-bobo-fay_n_4234898.

Norris, Maggie A., and Donna Rae Siegfried. "Taxonomy of Homo Sapiens." *Dummies*. See www.dummies.com/education/science/anatomy /taxonomy-homo-sapiens.

"Orang Pendek: A Cryptozoological Animal in the Sumatran Wilderness."
Wild Sumatra. See www.wildsumatra.com/orang-pendek.

Patterson-Gimlin film. See en.wikipedia.org/wiki/Patterson%E2%80%93
Gimlin_film.

Sessions, Larry. "Seeing Things That Aren't There? It's Called Pareidolia."
EarthSky. November 25, 2020. See earthsky.org/human-world/seeing
-things-that-arent-there.

Smola, Travis. "5 Bigfoot Legends from Around the World That
May Actually Be Real." *Wide Open Spaces*. December 23, 2020.
See www.wideopenspaces.com/5-bigfoot-legends-from-around-the
-world-that-may-actually-be-real.

"Sources of DNA." *Biology Project: Human Biology*. See www.biology
.arizona.edu/human_bio/problem_sets/dna_forensics_2/06t.html.

"Steps of the Scientific Method." *Science Buddies*. See www.sciencebuddies
.org/science-fair-projects/science-fair/steps-of-the-scientific-method.

Wayman, Erin. "What's in a Name? Hominid Versus Hominin." *Smithsonian*,
November 16, 2011. See www.smithsonianmag.com/science-nature
/whats-in-a-name-hominid-versus-hominin-216054.

"What Is Anthropology?" *American Anthropological Association*.
See www.americananthro.org/AdvanceYourCareer/Content.aspx?Item
Number=2150.

Yowie. See en.wikipedia.org/wiki/Yowie.

ACKNOWLEDGMENTS

I want to thank the members of the Bigfoot community who shared their stories, answered my questions, and didn't leave me in the woods. Thank you to the scientists who gave me their time and willingly clarified their work, especially Dr. Ian Tattersall, Dr. Todd Disotell, and Dr. Jeff Meldrum. I'm so appreciative of Jen Weddle and Cari Haug, whose expertise in science education proved invaluable in making sure I had my facts straight. I also want to thank my lovely and ever optimistic agent, Laura Nolan, who embraced this idea wholeheartedly. Three cheers to the creative and fun team at ABRAMS Kids—publisher Andrew Smith, managing editor Amy Vreeland, production manager Kathy Lovisolo, design manager Chelsea Hunter, and especially, assistant editor Sara Sproull and editor Howard Reeves, who understood the magic of Bigfoot. Rafael Nobre made this story shine with his illustrations and design. Additional thanks to copyeditor Richard Slovak, who supported my em-dash habit. Much gratitude to Alicia Lincoln, for her eagle-sharp eyes, and special thanks to Kelsey Ray, for her infectious spirit of adventure. Enormous, Sasquatch-sized hugs go to Alison Schiffern, Beth Agnew, and Kira Appelhans for their long-distance love, advice, and sanity checks. I'm grateful to my sister, Ashley Krantz, for her support and to my parents, Chip and Louise—they encouraged me to read voraciously and to spend lots of time outside, both of which made it possible to tell this story. And finally, to my husband, Scott Carney, the driving force behind this book and a constant source of encouragement. He has enthusiastically accepted Bigfoot into our lives, embracing all the weirdness that comes with it.

INDEX

Note: Page numbers in *italics* refer to illustrations.